1 MONTH OF
FREE
READING

at

www.ForgottenBooks.com

By purchasing this book you are eligible for one month membership to ForgottenBooks.com, giving you unlimited access to our entire collection of over 1,000,000 titles via our web site and mobile apps.

To claim your free month visit:

www.forgottenbooks.com/free90167

ISBN 978-0-484-58564-4
PIBN 10090167

BIOGRAPHICAL MEMOIR

OF

ASAPH HALL

1829-1907

BY

GEORGE WILLIAM HILL

———

READ BEFORE THE NATIONAL ACADEMY OF SCIENCES
APRIL 23, 1908

———

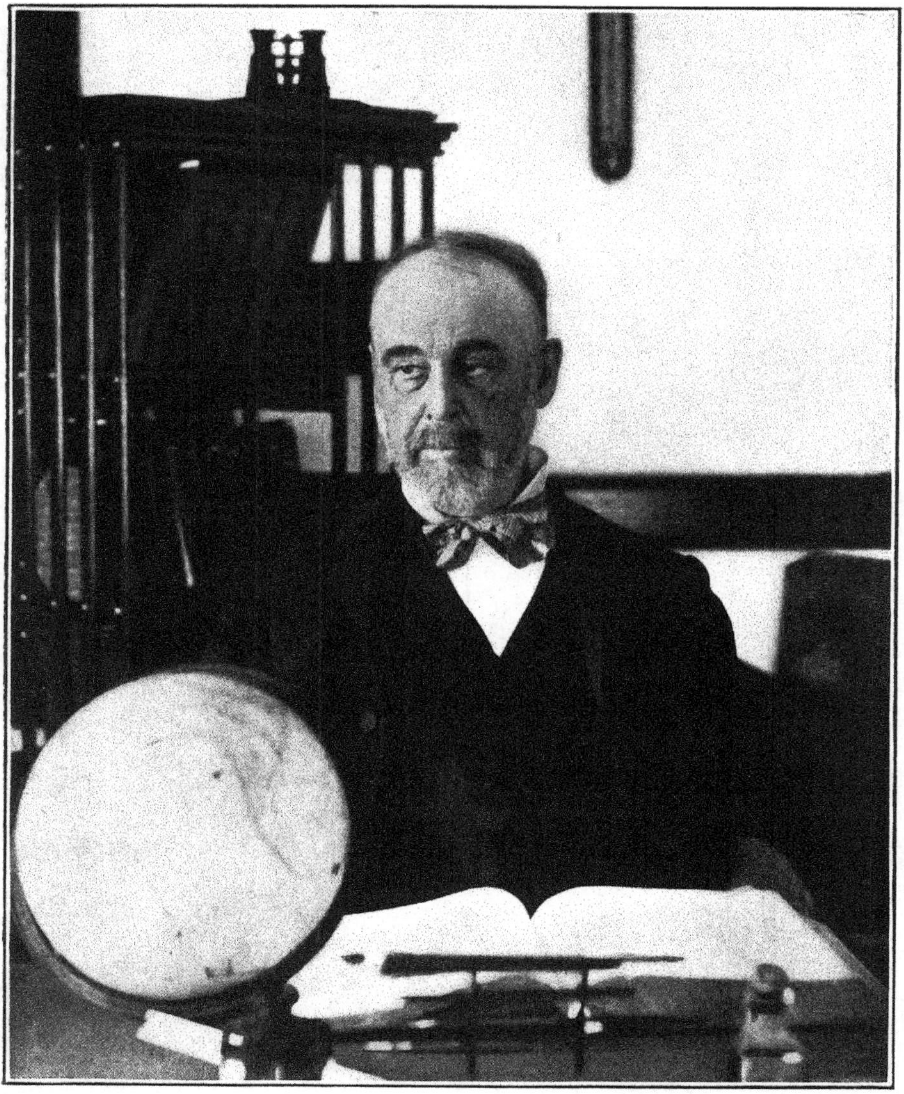

Yours, truly

Asaph Hall

BIOGRAPHICAL MEMOIR

OF

ASAPH HALL

1829-1907

BY

GEORGE WILLIAM HILL

READ BEFORE THE NATIONAL ACADEMY OF SCIENCES
APRIL 23, 1908

BIOGRAPHICAL MEMOIR OF ASAPH HALL.*·

In commencing the story of a remarkable man of science, it is necessary to say something of his lineage, in spite of the generally held opinion that the details of genealogy make dry reading.†

ASAPH HALL undoubtedly descended from John Hall, called of New Haven and Wallingford to distinguish him from the other numerous John Halls of early New England (Savage makes no less than seven before 1660), and who arrived at New Haven shortly after June 4, 1639, as he is one of the after-signers of the New Haven Planters' Covenant. His movements before his arrival are in some obscurity. From his son Thomas of Wallingford receiving a grant of fifty acres of land from the General Court of the Colony at the session of October, 1698, "In consideration of his father's services in the Pequot war," it is inferred that he was a dweller in the colony in 1637. At this date there were only four settlements in Connecticut, and it is supposed that the John Hall of New Haven is identical with a John Hall who appears as the holder of lots in Hartford about 1635. The genealogists are in dispute in the matter. Mr. Shepard sums up thus: John Hall came with the advance Hooker party, in 1632, or perhaps on the *Griffin* or the *Bird* (two vessels whose arrival at Massachusetts Bay is mentioned by Winthrop), September 4, 1633. He may have been in one or both of the overland exploring parties to the vicinity of Hartford, of which John Old-

---

* I desire to acknowledge the great assistance rendered by Mrs. Asaph Hall and the sons of Professor Hall in the composition of this memoir. Family documents in manuscript have been, in great abundance, put at my disposition. Dr. Horigan has contributed the elaborate bibliography.

† In the matter of the lineage of Professor Hall, besides the information supplied by the family, the following authorities have been consulted: Savage's Genealogical Dictionary; Davis's History of the Town of Wallingford; David B. Hall—the Halls of New England; Hibbard's History of the Town of Goshen; James Shepard—John Hall of Wallingford, a monograph. New Britain, 1902.

ham led the first; these took place in 1633 and 1634. He came with the advance settlers to Wethersfield in the fall of 1634, or to Hartford in the fall of 1635; went to the Pequot war in 1637; then, disposing of his lots by sale as early as the law allowed him to do without forfeiture, he removed to New Haven in 1639 or 1640.

From the record that he was "freed from training" in 1665, it is inferred that he was born about 1605. He married Jane Woolen, presumably at New Haven, although there is no record of the marriage. This lady had lived in the family of William Wilkes, and had been promised a portion of £10 on the occasion of her marriage, which promise was never fulfilled. Consequently, in 1647, Mr. John Hall sued the estate of Wilkes for the amount, which the court allowed. The records of New Haven give other facts in reference to John Hall, which may be read in Mr. Shepard's monograph.

In 1670 three of his sons—John, Samuel, and Thomas—decided to take part in the settlement of Wallingford. He himself, after a residence of about thirty years in New Haven, must have followed them shortly after. In 1675 he is selectman of Wallingford. He died not long before May 3, 1676, as at this date an inventory of his estate was taken. The widow, after marrying Mr. John Cooper, died some time in 1690.

The genealogists are not agreed as to the children of John Hall, but it seems there were seven—five boys and two girls. The fourth child was Thomas, baptized March 25, 1649; he was one of the proprietors of Wallingford in 1670; there he married, June 5, 1673, Grace (Watson). This is the first marriage recorded in Wallingford. His occupation is given in the land records as "carpenter." He died as "Sergeant" Thomas Hall, September 17, 1711.

The fourth child of Thomas Hall was Jonathan, born July 25, 1679, and died January 15, 1760. He married Dinah Andrews May 12, 1703, who was born July 25, 1684, and died October 24, 1763.

The first child of Jonathan Hall was David, born October 16, 1705. He married Alice Hale, September 23, 1730. He took part in the French and Indian war, and was killed in the murderous fighting at the south end of Lake George in 1755.

Up to this time the family line we are following resided in Wallingford; but now David Hall became an original proprietor in the town of Goshen, owning two rights. This town was organized in 1739. David's son Elkanah made the settlement as required, but on his return to Wallingford his brother Asaph took his place.

The fourth child of David Hall was Asaph, born June 11, 1735. He removed to Goshen before July 4, 1758, and became a prominent citizen of that town. In the Revolutionary war he was a first lieutenant, going to Ticonderoga with his company under Col. Ethan Allen; was in the second battalion, Gen. David Wooster's regiment, in 1777; was captain in Colonel Sheldon's regiment, Second State battalion, in service against Tryon's invasion. He represented his town in general court continuously for twenty-four sessions, beginning with that of 1773; he was also a member of the convention which adopted the Federal Constitution. He remained unmarried until nearly the end of his career, when, July 25, 1799, he married Esther McNair, a woman much younger than himself. After his military services were over, he was made a justice of the peace, and a great heavy armchair is still preserved in the family, seated in which he dispensed justice to the litigious inhabitants of Goshen. Unfortunately he fell from a sled and received injuries that proved fatal. He died about March, 1800. He left about a thousand acres of land, eight or ten houses, several sawmills, etc. This was the grandfather of the subject of this memoir.

His only son, Asaph Hall Second, was posthumous, being born August 8, 1800. His mother took, as her second husband, Seth Baldwin, and died in Goshen, May 7, 1851. He was prepared to enter Yale College, and passed examinations for the sophomore class when he was about 17, but for some reason did not go. His father's estate was turned over to him at the age of 19. He attempted the occupation of a merchant, but failed through the mismanagement of his property. To retrieve his affairs he established a clock factory at Hart Hollow, in Goshen, and manufactured clocks. He used to load up wagons with them and drive south as far as Georgia, selling the clocks and finally the horses and wagon, and return home for another trip. He married Hannah C., daughter of Robert Palmer. She was born at Goshen, August 19, 1804, and died there March 7, 1880.

These were the parents of Prof. Asaph Hall; they had six children, two boys and four girls, the Professor being the oldest. The latter was born October 15, 1829, in a house which has long since disappeared, situated on the road locally known as Goshen East street. This road for a distance of three miles passes over a plane region sloping upward towards Ivy mountain, nearly devoid of trees and wholly devoted to grass. This mountain is a Coast Survey station and has an elevation exceeding 1,600 feet, but it makes small impression of height, as the region surrounding the base is but 300 feet lower. The drainage is, on one side of the ridge, into the Naugatuck river; on the other side it passes through Bantam lake and the Shapaug into the Housatonic river. This is a bleak region at all times, but especially in winter, as there is nothing to check the course of the wind. At the time of Professor Hall's birth Goshen was a much livelier place than it is now, the population having fallen off 40 per cent in the interim. The town has been drained of its inhabitants by the attractions of factory life in the valley of the Naugatuck.

One of the earliest incidents of his life that Mr. Hall remembered was his father taking him to North pond on a summer day for a swim. He was placed upon his father's back, and then the latter struck out to cross the pond. The poor child, having never been in the water before, was in mortal terror lest he should be drowned. "Don't be frightened at the water, my son; only hang fast to papa's neck and you will be brought safely to land;" which was accomplished.*

Young Hall was, of course, sent to the "district school;" the building is still standing, although it is now used for quite different purposes than those of education. He was early initiated into the labors of farm life, beginning doubtless with driving cows to pasture.

---

* This pond is strangely omitted from the map of the State by the U. S. Geological Survey. There is no excuse for this, for, though hidden by swampy woods, it is visible from at least two points on the highway running on the west side. It has an area of about 200 acres.

This bucolic state of affairs lasted until he was 13, when his father, in one of his clock-selling trips to the South, was taken ill, and died at Clinton, Georgia, September 6, 1842. This was the beginning of hardship for Asaph. His father's land was dreadfully encumbered with mortgages. Nevertheless the family attempted to redeem at least one farm. A cheese factory seems to have been started (Goshen being an excellent grazing town). Professor Hall, writing to his son Angelo, October 20, 1906, speaking of his own mother, says: "She did a great deal of hard labor. One year we sold 10,000 pounds of cheese, all made by her." This seems incredible, and a suspicion arises that an extra zero has, through inadvertence, been added. Asaph, being the oldest of the children, must have been much driven in assisting his mother, since his only brother, Lyman, was several years his junior. In spite of this great effort, the main object failed of accomplishment; it was found that the utmost that could be done was to pay the interest on the mortgages and the living expenses. At this time, to help matters, Asaph carried the mail-bag on some route in Goshen.

As there seemed to be no hope of raising the mortgages, his mother retired to a small farm which, fortunately, she owned in her own right; and he himself, or his family for him, decided that he must learn the carpenter's trade. Accordingly he was apprenticed at the age of 16 to a carpenter, the apprenticeship to last for three years. In the summer of 1847, while he was still an apprentice, he was engaged with others in drawing timber from the Canaan mountains for buildings in the Blackberry valley, and noted with surprise that the people of that valley were able to raise good melons, while he in Goshen had always failed to bring them to maturity. The 700 feet difference of elevation was the principal cause of this. Whatever he could save from his scanty wages he turned over to his mother. At the expiration of his apprenticeship he set out as a journeyman. His reputation about Goshen was that of a skillful house-builder. He himself has informed me that he built houses as far away from his home as Great Barrington, in Massachusetts.

A prominent and well-to-do citizen of Goshen, known as Deacon Norton, who was the inventor of pineapple cheeses, employed him to make the wooden moulds for shaping them; these he made with chisel and mallet. The deacon, being a strict Sabbatarian, would say to him as he departed Saturday evening, "Now, young man, if you are reasonably quick in your motions, you can reach home before sunset."

During this period he was not without some advantages. His father having been a man of intelligence, books on a variety of subjects had accumulated about the house in much greater number than was the case with the neighbors. These he was diligent in reading at spare moments, after candles were lit or by daylight on Sundays. He was very eager to acquire information that might be serviceable in his future career. The bibliographer was innate in him. The first thing he did, on becoming interested in any subject, was to find out the names of all the prominent authors who had written on it, and the titles of their works, and this when there was small hope of his ever possessing or even seeing them. This, in after life, led him to despise the school and college text-book. "Intellectual pap, suitable only for babes," he would say. He always went to a master for instruction. When invited by the book-publisher to write a "Popular Astronomy," he made it a rule always to decline. "There is enough of such trash afloat without my adding to it."

About this time he is asserted to have entertained the ambition of exchanging carpentry for architecture. His predilection this way can hardly have been very decided, as in his after life I never heard him speak in praise of this art. As house-building was impracticable in winter, he had some leisure at this season, and managed to spend one winter at the Norfolk academy, studying algebra and six books of Euclid under the instruction of the principal.

After he attained his majority, doubtless with the approval of his mother, he began to lay up a sum of money that, in time, would suffice to pay his way through college. As his wages were only a dollar and a half a day, this was gaining way at a snail's pace. In the summer of 1854 he had accumulated about $300, and was becoming impatient to see his education begin. Having read in the *New York Tribune* a description of Central college, at McGrawville, Cortland county, New York, in which it was said that the price of tuition was remarkably low, and every student had an opportunity of paying his or her way through the college by manual labor, he thought that this was, perhaps, just the thing he wanted. After some reflection he determined to go.

This college had been founded by Gerrit Smith, the noted abolitionist of Peterborough, and kindred spirits. They had good intentions, doubtless, but the mode of advertising their advantages produced deleterious results. Their word was "Ho! every one that thirsted, come to the educational waters and drink, without money and without price; sex and color will be taken no note of." The consequence was the gathering of a motley crowd, many of them of the adventurer type, who believe that the sole use of education is that it enables one to dispense with muscular exertion. Among them an honest man would find little sympathy.

On his arrival, Mr. Hall was somewhat taken aback by the character of the place, so different from what he expected; but he decided to remain and make the most of his opportunities. There were no masters among the teaching corps, but he doubtless found some books in the library he was glad to be able to read, and the amount of his leisure greatly exceeded that he had previously enjoyed. Here it became evident that mathematics was the science to which he was preëminently attached. However, he did not, in his short stay, get beyond the elements; perhaps the institution afforded no means to this end. He acquired some French and Latin. His knowledge of carpentry stood him in good stead, for he kept the college buildings in repair during the whole of his stay. This was only a year and a half, when Mr. Hall became convinced that he had about exhausted the advantages at McGrawville, and that it would be useless to remain to take a degree. Moreover, he had become attached to a young lady attending the institution, Miss Chloe Angeline Stickney, of Rodman, Jefferson county, New York. She graduated in the summer of 1855, and, being two years ahead of Mr. Hall in the arranged course of studies, it so fell out that he was one of her pupils in mathematics. It is asserted that he, with other members of the class, tried to disconcert her by asking questions it was supposed she would not be able to answer; but they did not succeed.

At the end of 1855 Mr. Hall severed his connection with the college at McGrawville and came home. He was now engaged to Miss Stickney, and it was proposed that, after marriage, they should together undertake a school wherever an opening might be found. Miss Stickney had gone to Wisconsin to visit some relatives and in the hope that some school might be found in that State suitable to the wishes of herself and her affiancé. Mr. Hall joined her in February, 1856, and he also engaged in the quest for a school. These efforts were unsuccessful. Mr. Hall, who was a great hater of indecision, proposed that they should be married, and thereupon leave Wisconsin. Thus they were married at Elkhorn, Wisconsin, March 31, 1856, and started for Ann Arbor, Michigan. By this time Mr. Hall must have decided to be an astronomer.

Under date of April 2, 1856, Mrs. Hall writes to her sister Mary from Ann Arbor:

"Mr. Hall and I went to Elder Bright's and stayed over Sunday, and we were married Monday morning, and started for this place in the afternoon. Mr. Hall came here for the purpose of pursuing his studies. We have just got nicely settled. Shall remain here during the summer term, and, perhaps, three or four years."

Mr. Hall seems to have entered the sophomore class at the university, from which he passed to the junior at the next commencement. He studied French under Fasquelle and astronomy under Brünnow. For the latter he always entertained great regard as the teacher who had initiated him into the art of handling astronomical instruments. In this department of his science Mr. Hall was so apt a learner that he was at the bottom of an instrument in one-third of the time needed by the ordinary man. This struck Professor Brünnow so favorably that he was quite willing to give Mr. Hall extra attention. From this it came about that he passed out of the hands of Brünnow, after three months' instruction, well versed in the principles of observation. The practical training he could acquire by himself when opportunity offered.

Why he left Ann Arbor at this juncture is a little obscure; it could be induced only by the narrowness of his means. In order to get his degree, he would have to remain at the university two years longer, and he did not see where the money necessary for this was to come from. The couple left Ann Arbor in July, and went to visit a relative of Mrs. Hall at Hiram, Ohio. While here an opportunity offered of school teaching at the Shalersville Institute near by. This school was conducted by them from August, 1856, to the end of the following April. The remuneration was limited to the tuition fees of their pupils, but they seem to have given satisfaction, as a purse of $60 was presented to them by the townspeople at their departure.

During all this time Mr. Hall was burning the midnight oil, and here occurred the incidents of a characteristic anecdote which he himself narrated to me. After battling several evenings with an intricate problem in the motion of the heavenly bodies without success, he jumped up, much vexed at his incapacity, saying, "If I only could consult Laplace's *Mécanique Céleste* my difficulties would vanish in a moment. I dare say they have the book in the Western Reserve College library at Hudson. Tomorrow is Saturday, when 'school doesn't keep,' and I am going to see." Accordingly, after communicating his intention to his wife, in the early morning he set off on foot for Hudson, a distance of at least 15 miles, presented himself before the librarian, who, seeing a man in linsey-woolsey, his shoes covered with dust, supposed he was a tramp and was proceeding to turn him out without ceremony. But Mr. Hall, undaunted by his repulse, by a few words convinced the custodian of books that his poverty was far less than skin deep; the book was brought out, and Mr. Hall allowed to make all the extracts he wanted; after which he returned by the same method as he had come. At this time his library was so small that it could be packed in a corner of his trunk.

Mr. Hall was firmly set in his purpose to become an astronomer, and he saw that remaining in Shalersville would not promote it; so he determined to move once more. On the arrival of the couple at Cleveland they were still undecided which way to turn, whether to go back to Ann Arbor or make trial of fortune at Cambridge, Massachusetts; but shortly a heavy gale on Lake Erie settled the matter. Mrs. Hall was frightened at the great waves and shrunk from going aboard the steamer for Detroit. She said, "Let us go East." So she proceeded to her relatives in northern New York, and Mr. Hall to Cambridge, to see what opening there might be for them there. He conversed with the Harvard College professors, but the result was only partially encouraging. He could attend certain courses of lectures and receive help in his studies from the professors without any charge being made, but no salaried position could be guaranteed him; he would have to depend on stray scientific jobs. In this conjuncture Mr. Hall saw that it would be prudent to have a little money in his pocket before attempting settlement in Cambridge. Thus he went to Thomaston, Connecticut, and practiced again his trade of carpentry. A mill-owner in this place employed him to reconstruct the decayed apron of his dam across the Naugatuck river. At the completion of his task he was standing on the upper edge of the apron, contemplating his work with satisfaction, when, the timber being very wet from the spray of the fall, in a moment he lost his footing and slid the width of the apron splash into the pool below. The bystanders were indulging in a hearty laugh at his mishap; but the mill-owner said, "Don't mind them; you have made the best apron on the Naugatuck."

On May 19 Mr. Hall writes to his wife: "I get along very well with my work, and try to study a little in the evenings, but find it rather hard business after a day's labor. * * * I don't clearly know what we had better do, whether I had better keep on with my studies or not. It would be much pleasanter for you, I suppose, were I to give up the pursuit of my studies. I do not like to do this, for it seems to me that, with two years' more study, I could attain a position in which I could command a decent salary. Perhaps in less time I can pay my way at Cambridge either by teaching or by assisting in the observatory; but how and where we shall live during the two years is the difficulty. I shall try to make about $60 before the first of August. With this money I think I could stay at Cambridge one year, and might possibly find a situation, so that we might make our home there; but I think it is not best that we should both go to Cambridge with so little money, and run the risk of finding employment. You must come here and stay with our folks until I get something arranged at Cambridge, and then I hope we shall have a permanent home."

But Mrs. Hall insisted on sharing the lot of her husband. She indeed came to Goshen, but in six weeks the couple started together for Cambridge with the small capital of $50. On August 26 Mr. Hall entered the Lawrence Scientific School, and shortly began attendance on Prof. Benjamin Peirce's lectures. Prof. W. C. Bond had offered him $3 a week, and he had accepted the wages. Some of the devices for getting along were amusing enough. Mr. Hall found he must read the books of the German mathematicians and astronomers. Knowing the grammatical structure of the language, how was he ever to get sufficient leisure for learning the vocabulary? Mrs. Hall became a living dictionary, of which there were no leaves to be turned over, and from her Mr. Hall learned 30 or 40 new German words every morning as he kindled the fire. After some two months of this he thought himself tolerably equipped.

Early in 1858 he got some extra work—observing moon culminations in connection with Col. Joseph E. Johnston's army engineers surveying in the West. He received a dollar for each observed culmination, and in March he made 23 such observations. Mrs. Hall would awaken him out of his sound sleep in time to go to the observatory as faithfully as an alarm clock. He also eked out his small means by computing farmers' almanacs. During the latter part of 1858 he had other extra work—computing on the Colorado Survey. As to Professor Peirce's lectures, he soon ceased going to them, for he found they were too theoretical, and, moreover, there was some friction between the observatory and the mathematical department that rendered it unpleasant for him to be connected with both. Early in 1859 Mr. Hall's pay was increased to $400 a year. When Mr. Hall arrived in Cambridge, Mr. George P. Bond told him he would starve if he followed astronomy. This was not very courteous, as he himself was following astronomy at the time without starving. Mr. Hall might have retorted, "Why don't you follow your own advice?" Two of Mr. Hall's sisters visited him in Cambridge, and were so much dissatisfied with his poverty that they advised Mrs. Hall to persuade her husband to adopt some more profitable profession. Of course their advice was not followed.

Mr. Hall now began writing for the scientific journals; his earliest paper of importance is in the *Mathematical Monthly*, vol. III. It is on the transformation of an infinite series into a continued fraction. But most of his articles gave the elements of newly discovered comets and minor planets.

A few items respecting his occupations can be extracted from Mrs. Hall's letters.

May 4, 1858: "Another new comet last night. Mr. Hall has just finished computing the elements of one. They are to be published in the *Astronomical Journal."*

December 16, 1860: "Mr. Hall is up almost every night this winter observing zones."

July 18, 1861: * * * "Big Asaph computing orbits; this will make the sixth he has done since March, I think, besides computing two almanacs and writing a long article for the *Mathematical Monthly."* (Big Asaph is a playful designation of her husband; there was now a little Asaph in the family.)

In the summer of 1862 Mr. Hall's salary was again raised; but it was still inadequate to defray the expenses of generous living, especially as the Civil War had raised the prices of all commodities. Hence, when Congress authorized four new aids for the U. S. Naval Observatory, Mr. Hall, although reluctant to leave Cambridge, determined to apply for one of these positions. He accordingly went to Washington to undergo the examination, and returned to await the result. He learned in due time that he had passed and had been appointed. On August 6 he proceeded again to Washington to remain there permanently. Mrs. Hall and her boy remained at Cambridge for a few weeks, until the disturbance caused by the fighting about Washington should somewhat subside. She rejoined her husband just after the battle of Antietam. The rude alarms of the war prevented an easy flow of existence.

August 27, Mr. Hall writes to his wife, still in the North: "I am getting along as well as I can; have not looked for a place since my visit to Georgetown on Saturday; shall go there again next week. I think we shall like Georgetown better than Washington. The Observatory is not far from there, and now that a horse railroad is built between the two towns, communication is easy. I sometimes wish I could have stayed in Cambridge. It is much pleasanter there and quieter and better for study. But we will be content. We can be happy together almost anywhere."

The second battle of Bull Run had just been fought. From the Naval Observatory Mr. Hall heard the roar of cannon and rattle of musketry. In the following letter he assured Mrs. Hall that he was safe.

Washington, September 6: * * * "You must not give yourself any uneasiness about me. I shall keep along about my business. We are now observing the planet Mars in the morning. Mr. Ferguson and I work on alternate nights. You had better take your time and visit at your leisure now. Things will be more settled in a couple of weeks."

The battles in the vicinity of Washington filled the city with wounded soldiers, some of whom were relatives or friends of the family, and not a little time was spent by Mr. Hall in looking after the latter. Some were brought to the house in which he was temporarily living, and nursed there by his wife and himself. He had a severe experience in getting acclimated to Washington, so much so that he wrote to me to exchange places with him, in a joking way, of course, as he knew the Navy Department could not allow such a thing. The military camps in and about the city produced a very unsanitary condition of things, and diphtheria and smallpox invaded the house in which he was staying, and in March, 1863, he was obliged to send his family North. He himself remained in the city through the summer. Frequent removals were necessary, and it was only after five years' stay that Mr. Hall became the owner of a permanent residence in Georgetown.

In the spring of 1863 Professor Hesse resigned; this left a vacant place in the corps of professors of mathematics in the navy. Mr. Hall's friends desired that he should be promoted to it; but he, believing that the office should seek the man, would do nothing but wait. His wife, however, thought this was too bad, and ventured to address a letter to Captain Gillis on the matter from Cambridge, Massachusetts, where she was then staying.

On May 3 Mr. Hall writes to Mrs. Hall: "Yesterday afternoon Captain Gillis told me to tell you that the best answer he could make to your letter is that hereafter you may address me as Professor Hall. * * * You wrote to Captain Gillis, did you? What did you write?" Captain Gillis was a kindred spirit with Professor Hall, and thus the latter's promotion appears to have taken place with little friction. Thus, after many discouraging circumstances, the battle for recognition was ended.

Some pleasant incidents occurred about this time. One evening Professor Hall showed President Lincoln and Secretary Stanton some objects with the equatorial. A few nights afterwards the President, unattended, repeated his visit; he wanted to know why the moon had appeared inverted in the telescope; and when his wishes on this point had been gratified, he remained to converse generally on astronomy.

On the morning of July 12, 1864, firing was heard north of the city; General Early was threatening Washington from that side. Professor Hall went to his work as usual, but he did not return. Mrs. Hall, with her little boy, set out for the Observatory to ascertain what was become of him. A note on his table explained his absence. "I am going out to Fort Lincoln; don't know how long I shall stay; am to be under Admiral Goldsborough. We all go. Keep cool." Together with the other officials at the Observatory, Professor Hall was put in command of a number of workmen from the Navy Yard who manned an intrenchment near Fort Lincoln. Many of them were foreigners, and some of them did not know how to load a gun. However, the Observatory and Navy Yard people did not have to endure the strain of this novel position a long time, for the Sixth Corps had early been telegraphed for, and they now came up the river to relieve them.

Having now traced Professor Hall through his attempts to get a footing in the scientific arena, we are at leisure to note his contributions to the science of astronomy.

In 1863 he deduced a value for the constant of solar parallax from observations of Mars with equatorials at the observatories of Washington, Upsala, and Santiago during the opposition of 1862. As a uniform method of observing was not followed at the three observatories, the result 8."84 has not, perhaps, a high degree of precision. Some investigations made by Professor Hall afterwards showed that the method with two micrometer wires cutting off nearly equal small segments of the planet's disk is better than with a single tangent wire.

In 1864 Professor Hall, casting about for something to do with the instrumental means at his command, determined to form a catalogue of the stars in the cluster Præsepe. Eleven of the brighter stars were determined by the meridian instruments of the Observatory, and the remainder were interpolated by means of differential measures made at the Equatorial. The number of stars determined was 151. This work stretched from 1864 to 1870. It is as good as the appliances employed permitted. Better results could, doubtless, have been obtained by the use of a heliometer.

In 1868, May 2, Professor Hall observed an occultation of Aldebaran in unusual circumstances, the moon being invisible at the time and only 8° distant from the sun (Washington Observations for 1868, p. 327).

The same year an article on the "Positions of the Fundamental Stars" was contributed to the *Astronomische Nachricten* (vol. LXXI, p. 191), in which the haphazard methods of observing in vogue at that time were criticised and some improvements suggested. Some of the latter have since been adopted.

In May, 1869, Professor Hall was ordered to observe the solar eclipse of August 7 on the eastern coast of Siberia. He proceeded to San Francisco by way of the Isthmus, and thence by the United States vessel *Mohican* to Plover bay, his destination, where he arrived July 30. The eclipse was observed, but the data obtained were of no great importance. This was due partly to clouds and partly to insufficient instrumental outfit.

Professor Hall says in his report: "With regard to the eclipse, I ·most sincerely regret that we had no means of taking photographs. As the weather happened to be, this was the one thing most needful for us, and I hope that our fortune in this respect may be a warning to future expeditions."

The following year Professor Hall observed the solar eclipse of December 22, at Syracuse, in Sicily. He spent the short time of totality in noting the appearance of the solar corona, and indicates his disbelief in the scientific value of hand-drawn pictures. The utmost one can do is to note the general impression produced. Deliberation in sketching an outline is impossible. In this respect it is much like the aurora borealis.

The secular perturbations of the eight major planets of the solar system was a subject that at all times interested Professor Hall. He was employed by the Smithsonian Institution to report on Mr. Stockwell's investigation. This he did in his accustomed thorough way, testing his figures in many places by his own computation. This led him in 1870 to publish an article in the *American Journal of Science* noticing the imperfections of previous treatments of this matter. He shows how important it is we should start with values of the masses quite approximate. However, it is still more important that the squares and product of the masses of the large planets, Jupiter and Saturn, should be properly taken into account in the investigation.

In 1872 he reported on observations of Encke's comet, both for position and physical appearance, during its apparition of 1871. Sixteen observations for position were obtained between November 2 and December 7.

Professor Hall says: "I first saw this comet in 1858, and have observed it at four returns. My impression is that it has been fainter during its present return than I have ever seen it before, considering its distance from the earth and sun. To an observer who sees such changes going on, the question will naturally occur whether his determinations of position made at different times and when the comet has such different forms are strictly comparable. It would appear that this should be carefully considered in any discussion of the motion of the comet."

In connection with this matter, Professor Hall, about this time, contributed to the *American Journal of Science* an article on the "Astronomical Proof of a Resisting Medium in Space." He considers that the comets of Faye and Winnecke do not support the hypothesis of a resisting medium, which Encke thought thoroughly established by the motions of his comet. Professor Hall points out the logical incompleteness of the proof of the action of a cause simply because it accounts for the observed facts; it is necessary to show, in addition, that no other cause is capable of so doing. His prevision in this matter was amply confirmed by the subsequent researches of Von Asten, when it was found that Encke's modulus of resistance had to be greatly cut down.

Professor Hall was the chief of the party sent to Vladivostok to observe the transit of Venus of December 8, 1874. Six men accompanied him in various capacities. He left Washington July 11 and arrived at San Francisco July 21. Messrs. Harrison and Gardner, in charge of the instruments of all the northern parties, made the journey from Washington to San Francisco on freight trains. The party sailed from San Francisco July 28, in the steamship *Alaska,* and reached Yokohama August 20. Leaving this place on the 26th, they reached Nagasaki August 30. On September 3, aboard the U. S. steamer *Kearsarge,* the whole party was conveyed to Vladivostok, which was reached September 7, and baggage and instruments were landed September 9.

A battle was immediately commenced with the untoward climate of Vladivostok. The party seem to have been at their wits' end how to protect their instruments and still leave it possible to observe with them. Driven away from their first chosen site by the wind, with difficulty they found a second. Professor Hall gives a somewhat sarcastic account of the conditions: "It was a puzzling question what to do with the equatorial house. The winds at Vladivostok are extremely violent, and the roof of this house was so designed as to give the winds a strong hold of it. Finally, four posts were set firmly in the ground; holes were bored in the roof of the house; strong ropes passed around the main rafters and made fast to the posts. To keep the wind out of the house, a curtain was made of old canvas, furnished by Captain Harmony, and nailed and drawn down around the body of the house. In this way the roof was made secure, and wind and snow were, in a good degree, kept out of the house. But, at the same time, the house was rendered nearly useless for ordinary observations, since if we untied the fastenings and opened the shutters we were in danger of having the roof blown off and our equatorial broken. Afterward I regretted not having followed the plan proposed by Captain Harmony and by Mr. Smith, of building a strong plank fence a few feet outside the house, bracing this fence from the inside and banking it with dirt on the outside, and letting the planks rise a few feet above the eaves of the roof; but I had such an experience in building that I very much disliked undertaking anything more in Vladivostok."

The transit instrument was found to have been very slovenly constructed. Professor Hall says: "Of the three spirit-levels furnished by the maker, two were utterly worthless, and the third was kept to be used only in case of necessity."

With regard to the photographic apparatus, he says: "The heliostat and clock had been so profusely oiled that they would not run in cold weather. They were carefully cleaned by Mr. Gardner; no oil was applied except a little at the bearing of the pendulum, and afterward they performed tolerably well. * * * In accordance with the last directions received from the commission, we had covered our heliostat pier and the coffin-shaped plate with white cloth. After this was done the photographs were blurred, and it was at first supposed that the focal distance had been erroneously measured. The coffin-shaped plate was moved and the focal distance changed to correspond with that found by experiment with ground glass, but the photographs were no better. After some delay it was found that the trouble came chiefly from light reflected from the white cloth and other surfaces to the sensitive plate. The cloth was removed and the surfaces painted a dead black color, with lamp-black and water. There still remained a source of trouble which it was impossible to avoid. This was the difference of temperatures inside and outside of the photographic house, and which varied from 20° to 50°. This allied itself with what seems to me the weakest part in the excellently designed apparatus provided by the commission for doing the photographic work—that is, with the fact that the glass mirror reflects but a small percentage of the light. The reflection of so small a quantity of light through the photographic lens made it necessary to open the slit in the slide to its greatest width, 2½ inches, and even then, at low altitudes to move the slide by hand slowly past the plate, and thus lengthen the time of exposure. But if the time of exposure was much lengthened, the difference of temperatures made the photographs blurred and indistinct around the edge of the sun. It was decided to make faint and sharp photographs rather than those with blurred edges, and as dense as the photographers had been instructed to make."

"On the morning of the day of transit the apparatus was in good adjustment, and Mr. Clark and his assistants had everything in readiness for making 200 photographs. At the time of the first and second contacts the haze prevented them from making any photographs, although the telescopes gave us faint but very fine images of the sun and Venus. After Venus had entered wholly on the sun's disk the haze became thinner, and a few photographs were made, when we were stopped by denser haze. About an hour after this another set of photographs was made, and still later, and near the end of the transit, nine photographs were made. There are 13, I think, that will admit of measurement. The others will be found probably too faint to be used."

Thus, on account of atmospheric haze and imperfect apparatus, the expectations entertained were not fulfilled.*

In 1866 and 1867 Professor Hall was assistant to Professor Newcomb on the new transit circle. For a few months at the end of 1867 he was in charge of this instrument. From 1868 to his departure for Vladivostok he was in charge of the small equatorial. In May, 1875, he was put in charge of the Clark refractor, which he held to his retirement.

In 1875 he proposed to derive the mass of Mars from the motions of the minor planets, whose periods are nearly double that of the first-mentioned planet. He gives the formulas for deriving the coefficients of the long-period terms in the longitude of the minor planet, and a table to facilitate the computation, and then makes application to 32 of these bodies. The practical objection to this is that we have to wait until a large fraction of the period has elapsed.

On December 7, 1876, Professor Hall noticed a brilliant white spot on the ball of Saturn. It continued to be visible until January 2, 1877. He immediately made use of it to determine the time of rotation of Saturn. Sir William Herschel had made a determination, but the statement of it in the books had got vitiated.

In a memoir of the same year (1877) he considers the equation of the curve of the outline of the shadow of the ball of

---

* I have not been able to find Professor Hall's report printed in any scientific journal or collection. Fortunately, what appears to be an unabridged edition of it is published in the *New York Daily Tribune* for March 26, 1875.

Saturn on the ring. The treatment of this question is greatly simplified by the consideration that the diameter of the sun as seen from Saturn is only about 3'; hence, for all practical purposes, it suffices to reduce that body to a point. But Professor Hall treats the question in its most general form, and the equations evolved are interesting from the mathematical side.

We now come to the principal achievement of Professor Hall, the discovery of the two satellites of Mars. The exploit created so much excitement in the scientific world at the time that even now, after the lapse of thirty years, the history of it seems very trite. Of course, when men have made a long and painful search for a thing without success, they feel warranted in saying it does not exist. In this way the text-books repeated the statement, "Mars has no moon." But the opposition of Mars in 1877 was a very favorable one, and, still more, the instrument Professor Hall had at command considerably exceeded in space-penetrating power any that had been previously constructed. On these accounts he thought that it could not be laid to his charge that he had wasted his energies, even if he discovered nothing, for negative knowledge is sometimes as valuable as positive.

Mars reached its stationary point August 5, and a month later it was in opposition. The search was begun early in August. Between the 4th and the 10th no observations are recorded, but on the 10th Mars' white spot is observed, with the remarks: "Planet blazing; clouds passing." On the 11th the remark is: "Seeing good for Mars." At $14^h 42^m$ a faint star near Mars was measured (this was Deimos), when fog from the Potomac interrupted observation. Unfavorable weather continued to August 15, when the white spot of Mars is seen, and the remark is: "Images blazing." On the 16th Deimos is seen and measured twice, the result showing that it was attending Mars. On the 17th not only is Deimos seen, but a new object is detected. To the later observation is added the remark: "Daylight. Both the above objects faint, but distinctly seen by George Anderson and myself." The observations of the 17th and 18th clearly established the character of these objects, and Admiral Rodgers publicly announced the discovery. For some time the inner moon was a puzzle, but by making a prolonged watch on the nights of August 20 and 21 the identity of the object at its several appearances was made out. The glare of Mars so generally interfered with the visibility of the satellites that it was necessary to slide the eye-piece so as to keep the planet outside the field of view.

This discovery produced a great sensation. I remember visiting my bookseller in New York just as the public announcement was made. The first thing he did on my entering his place was to thrust the morning paper under my eye and point to an editorial, with "What do you think of that?" I took time to read the heading and the first line, and replied, "That's a moon hoax." Information came from Washington the next day that effectually disposed of this view. A craze for discovering satellites took possession of many people. One man in Pennsylvania, the owner of a small telescope, wrote to Professor Hall, "I have discovered a satellite, and it goes round the primary in five seconds." Envy, too, was not wanting; but Professor Hall could afford to stand by and smile; the joke was all on his side.

As to the amount of credit due to Professor Hall in this matter, it is certainly a singular fact that after the discovery no astronomer put in a claim that he also had intended to prosecute a search for satellites, but had not yet taken opportunity to commence work. Thus it would seem that Professor Hall in 1877 was the only astronomer who indulged such an intention and carried it out. We shall do no one a wrong in saying that if it had not been for Professor Hall the great opportunity of 1877 would have passed without the discovery of these bodies. Professor Hall was an exceedingly modest man in reference to his achievement—modest beyond what the occasion called for. He more than once said to me, "Mr. George P. Bond should have discovered these satellites in 1862; his telescope was powerful enough to reveal them." I would reply in a bantering way, "He very kindly left the nuggets for you to pick up." After the discovery, it was surprising what a number of relatively small telescopes revealed these satellites; so much easier is it to discover what has already been discovered.

Measurement of these satellites continued to be the care of Professor Hall until he closed his work at the U. S. Naval Observatory, in 1892.

Professor Hall reported on the transit of Mercury of May 6, 1878, which he attended to in Washington. Mr. Joseph A.

Rogers having made 216 dry plates, 72 of them were exposed in Washington. The third and fourth contacts were observed; the others were missed. The diameter of the planet was measured with the double-image micrometer, but the results were deemed of slight importance.

The same year he journeyed to Colorado to observe the solar eclipse of July 29. His station was La Junta, on the line of the Atchison, Topeka and Santa Fé railroad.

Professor Hall reports: "The day at La Junta was all that could be desired for observing a total eclipse. My own special work during totality was searching for an intra-mercurial planet, the supposed Vulcan, indicated by the researches of Leverrier on the orbit of Mercury. Before the eclipse I studied the configuration of the stars as they are laid down on the chart published by the Observatory, and during totality a copy of this chart was placed a few feet in front of me, so that I could refer to it instantly. As soon as totality began I turned my shade to the free opening and commenced sweeping above the sun and near the ecliptic. My sweeps extended from the brighter part of the corona to a distance of about ten degrees from the sun. The magnifying power was so great that the sweeping could not be done very rapidly. In this part of the sky I saw nothing but the stars laid down on the chart. * * * The interior part of the corona seemed to me very bright—much brighter than the corona I saw at Syracuse, Sicily, in 1870. * * * The darkness during totality at La Junta was very slight, and I could read my chronometer and could see the stars on the chart with ease and without the aid of artificial light.

"At La Junta I was struck with the clearness and beauty of the sky at night. I could see distinctly and steadily with the naked eye λ Ursæ Minoris, a star that I have never been able to see at Washington in this manner, although I have made many trials and know, of course, just where to look. The number of stars visible to the naked eye was greatly increased. But the most striking change was in the appearance of the Milky Way. The outlines and divisions of this great starry ring were seen with wonderful distinctness. A few hours after sunset, when the intense heat of the ground had passed away, the images of stars in our telescopes were good. I cannot but think that those elevated plains offer advantages for astronomical observations that have not hitherto been made use of. An obvious objection to the use of large telescopes in those regions is the great change of temperature from day to night. It might be necessary that the observer should work from midnight until morning, but these are generally the best hours of the night at any place. This interesting question will not, I think; be definitely settled until a complete and careful experiment has been made. The reports of observers who have had an experience of a few days or weeks are apt to be conflicting and unsatisfactory. The elevated plains of our country are now easy of access, and in order to decide the question, an experienced astronomer with a good telescope should be placed in an elevated position, and should continue his observations for several years."

As more than one astronomer, on this occasion, announced the discovery of intramercurial planets, Professor Hall's remarks have an interest.

In 1880 Professor Hall gathered together and published his observations, 1,614 in number, of double stars. They include a few made in 1863 with the smaller equatorial. In the introduction, which is an admirable exposition of the difficulties which beset the observer with an equatorial, Professor Hall is very frank in stating the imperfections of the driving clock and the dome as well as the imperfection of his eyes.

In 1882 was published an investigation of the parallaxes of α Lyræ and 61 Cygni from observations with the large equatorial. A sensible temperature coefficient was found for the revolution of the micrometer screw, although previously it had been assumed that it was insensible. An abbreviated statement of the observations of α Lyræ made at the prime vertical from 1862 to 1867 is appended, showing a negative parallax.

In the autumn of 1879 the satellites of Mars were again observed. Professor Hall says: "The observations were made in the following manner: In order to avoid sliding the eye-piece, as was done in 1877, a piece of colored glass, covering one-half the field of view, was inserted in the forward end of the eye-piece, near the micrometer wires. It might be better to silver one-half the forward lens of the eye-piece, but an attempt to do this did not give a good result. In making the observations, the planet was placed behind the colored glass, through which the wire could be seen, and, both objects being kept near the center of the field, the angle of position and the distance were measured by bisecting the disk of the planet and the satellite. In this way the observations were made in much less time than by sliding the eye-piece." Both satellites were remarkably near the predicted position, and the corrections of the elements of their orbits deduced from this series of observations were quite minute.

Professor Hall was chief of the party that was sent to San Antonio, Texas, to observe the transit of Venus of December 6, 1882. He was measurably successful and brought away a considerable number of photographs. I have not been able to find his report in print.

In 1884 he discussed the observations of Hyperion. From 1852, when Lassell had observed this satellite of Saturn, up to 1875, when Professor Hall commenced to observe it, no observations had been made. The discussion brought out the interesting facts of a nearly constant eccentricity and a peri-saturnium retrograding about 20° a year. Professor Hall gave expressions for the perturbations arising from the action of Titan. These data soon led astronomers to perceive that Hyperion afforded an example of a periodic solution in the problem of three bodies. There was still some debate as to the mass of Titan; but this was soon settled.

His being in charge of the Clark refractor seemed to point him out as the man who should have the care of all the satellites of the solar system, excepting those of the earth and Jupiter. Hence in 1885 he published a discussion of his observations of Oberon and Titania, reaching from 1875 to 1884, deducing corrections to the elements and the mass of Uranus.

The satellite of Neptune was treated in a similar manner in another memoir composed about the same time.

Next Iapetus was investigated, the longer time this satellite has been known and the slowness of its motion demanding more extensive treatment. The memoir containing this investigation is among the most admirable pieces of astronomical literature. Nothing can exceed the clearness and precision of its statements, both on the observational and mathematical sides. We are reminded of Bessel. After the lapse of a quarter of a century, it is scarcely superseded. The amount of pains taken to have everything right is a subject for wonder. The elements are deduced from ten years' observations, extending from 1874 to 1884, except that observations of Sir William and Sir John Herschel were used in getting the period. The mass of Titan is assumed at a 10,000th part of that of Saturn; it is now known that it is more than double this. The mass of Saturn, derived by Professor Hall from his measures of the distance between the two bodies, is half a per cent greater than is indicated by the action of the planet on Jupiter. It will be noticed that the observations of 1874, 1883, and 1884 give a result that is quite different from that of the years 1875-1881. Some abnormal appearance of Saturn and its ring must have interfered with the measurement of the distance between the two bodies. Tables for the motion of the satellite conclude the memoir.

When the memoir on Iapetus was finished Professor Hall immediately took up the six inner satellites of Saturn and treated them on a plan precisely similar to that followed in the preceding memoir. Here he found a mass of Saturn still larger than that just mentioned; and a similar cause must have acted to vitiate the measurements of distance.

This was followed by the memoir entitled "Observations for Stellar Parallax," in which Professor Hall, after detailing his efforts to obtain a more certain value of a revolution of the micrometer screw, gives and discusses his observations for the parallax of two stars and rectified his previous discussion of the parallaxes of α Lyræ and 61 Cygni, it having been discovered that the correction for temperature had been applied with the wrong sign. For the latter stars some new observations were added to the discussion.

In 1889 Professor Hall gathered together the various notes he had made on the appearance of Saturn and its ring in the interval 1875–1889. The memoir contains also the measures he had made of the dimensions of the ring and their reduction. But the principal statement is the showing that the theory of Otto Struve, asserting the imminent precipitation of the ring upon the ball, was without warrant.

During these years Professor Hall was much engaged with the matter of removing the U. S. Naval Observatory to a new site.

This question was agitated for a long time before Congress authorized the removal and appropriated money for purchasing a new site. The unhealthfulness of the locality occupied by the Observatory was conceded by all who were in a position to know. This, however, was not sufficient to overcome the inertia; but at length it was seen that the Observatory would probably be in the way of the schemes for improving the water front of Washington. Thus Congress was induced to provide the sum necessary. A large number of places were offered to the Government for the purpose, and a board composed of Observatory officials was appointed to ascertain the advantages of each place and to report thereon. Professor Hall was a prominent man on this board. A very thorough investigation was made, and the present site of the Observatory (then known as the Barber place) was recommended. This so offended the unsuccessful offerers that they accused the board of undue partiality. The Navy Department thought best to appoint a new board, composed of scientists at a distance from Washington; but the new board simply confirmed the recommendation of the first. After the site was purchased, there was a great deal of friction in getting the new buildings erected. The first contract, after the time had been lengthened, was declared forfeited and a new one was formed. It was not till 1892 that the instruments were moved into the new quarters.

From the autumn of 1885 to the spring of 1887 Professor Hall measured the positions of 63 small stars, relative to the brighter ones in their neighborhood, of the group of the Pleiades. This was done in order that in the future it might be determined whether the two classes of stars had a general relative proper motion. The average magnitude of the small stars was 12.4.

About the same time Professor Hall was observing to get the stellar parallax of α Tauri. His result was only one-fifth of that of Otto Struve.

The next memoir of Professor Hall is on the constant of aberration. In this he gathers and discusses the equations resulting from the observations of α Lyræ made at Washington with the prime vertical instrument in the years 1862-1867. Some suspicion attached to these observations because they give a negative parallax for the star; but Professor Hall thought that the value of the constant of aberration derived from them was not greatly vitiated by this circumstance, and he declares his impression that the method of getting the constant of solar parallax from that of aberration is the best astronomers now have.

In a memoir on the extension of the law of gravitation to stellar systems, Professor Hall expresses the opinion that we are hardly warranted at present in making the extension.

As Professor Hall was to be retired on October 15, 1891, having then attained the age of 62, just before retirement he collected his observations of double stars made during the interval 1880-1891.

In the year 1889 the orbit plane of Iapetus returned to a position similar to that of 1875, except that the apparent motion of the satellite was reversed; hence in this year Professor Hall began another series of observations which were continued in the opposition of 1890. The principal result of the discussion of these was a smaller value of the mass of Saturn than was obtained in the earlier investigation, but still larger than the commonly accepted value.

In a memoir of 1892 on the relative motion of 61 Cygni, Professor Hall discusses his own and his predecessors' observations, with the result that the two components are physically connected. The addition of terms proportional to the square of the time brings the theory in accord with observation.

In a supplement to his second collection of observations of double stars, Professor Hall derives formulas representing the coordinates of the more interesting.

Although Professor Hall had been retired, the Superintendent of the Observatory tendered him the use of the Clark refractor for observing the satellites of Mars in the summer of 1892. The discussion of the observations is found in the *Astronomical Journal,* vol. XII.

Having, in the just-mentioned memoir, deduced a new value for the revolution of the micrometer screw, Professor Hall, in another article on the masses of Mars, Saturn, Uranus, and Neptune, modified his previously given values of these masses to accord therewith.

In a memoir of 1893 the perturbations of Flora by Mars and the earth are somewhat rudely derived, but with sufficient detail to show that they are of no great moment.

In "A Suggestion in the Theory of Mercury," a memoir of 1894, Professor Hall, using a formula of Bertrand, showed that the unaccounted-for part of the motion of the perihelion of Mercury could be explained if the exponent —2 of gravitation was algebraically diminished by a very small fraction. This suggestion was adopted by Professor Newcomb in forming his tables of the interior planets.

The same year Professor Hall published a memoir on the orbits of double stars, giving a very lucid exposition of the formulas and methods for deriving them from the observations.

In 1896 Professor Hall published the absolute perturbations of Nemausa of the first order by Jupiter. They are compared with normals covering 34 years of observation, and the accord is so good that tables might be constructed from this investigation accurate enough to give fair positions of this planet in the future.

In 1898 Professor Hall discussed the observations of the satellite of Neptune made during the preceding opposition, at the Yerkes Observatory, by Professor Barnard, and derived a new set of elements. He points out the desirability of continuing observations of this sort in order that the ellipticity of the planet may be determined.

Two years later Professor Hall, then residing at Cambridge, used his opportunity to investigate, as far as the preserved documents permitted, the observations of this satellite made by the two Bonds in 1847 and 1848. The conclusion he comes to is that the methods of treatment adopted for the Cambridge observations is quite superior to that followed at Pulkova.

In 1902 Professor Hall has a memoir considering the mass of the rings of Saturn. He first assumes plausible values for the elements of the problem, which lead to a negative value for the mass under consideration. However, it is easy to make such changes in the elements as will render the mass positive. The nebular hypothesis would lead us to suppose that the mass of the rings relatively to the mass of Saturn is not greater than the mass of the interior planets of the solar system relatively to that of the sun.

The last article Professor Hall contributed to the scientific journals is on stellar parallax, and bears date September 20, 1906.

From the preceding exposition it will be seen how varied and numerous were Professor Hall's investigations in astronomy in the course of little less than half a century.

A few details must be added concerning Professor Hall's social life after he was retired from active service in the United States Navy. In July, 1892, Professor Hall had to endure the sorrow of losing his wife. This lady, of great refinement and culture, but frail physique, had by her encouragement, sympathy, and self-sacrifice been of great service to her husband in the pursuit of his science. Having finished his work at the Observatory at the end of the summer of 1892, Professor Hall still lingered at his residence in Georgetown, frequently resorting to the Observatory library for scientific reading; but in the summer of 1894 he began to think that the ties binding him to Washington were of the slightest, and to indulge a wish to pass the rest of his days in the midst of rusticity. This led him to think of his native town, Goshen, where most of his relatives resided. Proceeding thither, he purchased a small piece of land with a house. The latter had been built shortly before the Revolutionary War and had served as a tavern, at which a line of stages, running to and from Albany, stopped in the period following this war. In order to be fit to live in, this dwelling had to be renovated, and Professor Hall spent the autumn and the following spring in attending to this work. In taking up the flooring some coins of colonial Connecticut were discovered. This dwelling stands on the road between Litchfield and Norfolk, about 200 yards south of the crossing of it by the Gunstock brook, the outlet of North pond. This is a very retired spot, the nearest railroad station, at Norfolk, being seven miles distant. The screech of the locomotive is not heard or only at rare times, and then very faintly. Telegraph or telephone poles are not visible on the road. This place is about two miles north of where Professor Hall was born. He dubbed it "Gunstock."

In 1896 he received an invitation from Harvard University to give instruction in celestial mechanics. Acceptance of this would involve an interruption in his plan of rustic life; nevertheless he decided to accept; but he stayed in Cambridge only during term time, all his vacations being spent at "Gunstock." At first his position at Harvard did not make him a member of the faculty; but it was soon discovered that he was a capital man in the social line, and his rank was so changed that he became a member of the governing board. His lectures were addressed to undergraduates and made an elective course for the obtaining the bachelor's degree. Professor Hall therefore could not indulge in research work before his class. As Gauss' *Theoria Motûs* mainly turns on spherical trigonometry, in which the students coming to him should be proficient, he adopted this book as outlining the drift of his teaching. After five years of this occupation, being now 72, he thought himself warranted in retiring to his rustic home.

In the autumn of 1901 he married Miss Mary B. Gauthier, who survives him.

Planing off the summit of a large rock near his door, he bolted to it an iron sun dial constructed after his own plan. He had some difficulty in getting the foundrymen in Torrington to follow precisely his drawing. The light of the sun was compelled to go through a conical perforation in the style ending in a pinhole. This formed a minute star on the horizontal plane of the dial which had a deep groove on the line of the meridian. The bisection of the star by the sharp line at the bottom of the groove marked the time the sun's center was on the meridian. By this device Professor Hall thought he got time correct to within 20 seconds. The dial was kept painted, and when not in use was covered by a wooden box. When the neighbors found out that Professor Hall had good time, they came to the house to set their watches.

To get the geographical position of his house, Professor Hall made a rude triangulation connecting it with the summit of Ivy mountain.

Almost the last episode of Professor Hall's career was a deputation of his fellow-townsmen waiting on him to see if he would not represent them in the Constitutional Convention, since he was by far the most notable man in the town; but his reply to them was, "Gentlemen, I think you do an injustice to your fellow-citizens who have the misfortune to live in the populous towns. If I go to the convention, I must vote for the reform of this abuse." The deputation, with downcast countenances, went off to find a candidate elsewhere, wondering how a man could be so quixotically altruistic.

About January, 1907, Professor Hall found himself incapable of performing the domestic labors about his house. One day he said to his wife, "I can't swing the axe; what do you suppose is the matter with me?" His friends were alarmed about his condition, and a medical examination showed heart trouble. Forbidden to exert himself, he was still able to walk about the house and grounds and ride to the neighboring villages. He spent the early spring at Annapolis, Maryland, where his son Angelo holds a position in the U. S. Naval Academy. Returning, he spent the summer at home; but at the beginning of November it was deemed prudent to remove him from the rigors of the Goshen winter. He seems to have endured the fatigues of the journey to Annapolis tolerably well, but he had not been there quite three weeks when the end came. He died November 22, 1907.

In noting the character of Professor Hall as a scientist and as a man, we observe that he has been called a self-made man, and this partly in commendation, partly in commiseration. But this statement is misleading, if it is supposed that it is equivalent to the fact that he had no continuous academic training. Whatever his surroundings, he always grasped any tool within his reach that would serve his purpose, and wielded it vigorously. If he had been wealthy enough to have employed a coach, I am afraid the coach would soon have been left behind, such enormous strides he was accustomed to take. Mr. Hall was a bird of the species that can run as soon as it is out of the shell. His battle was rather with poverty than lack of instructors. Speaking to me of his great admiration for the histories of Gibbon and Hume, he said: "What a fortunate circumstance it was for me that these books were in my father's library." He had actually read them as soon as he could read at all, and with a comprehension and appreciation equal to that of a man of forty. Perhaps he was the only boy of his age in Connecticut who indulged in that sort of reading. Dickens' Child's History of England was thrown away in his case. In a letter to me of February 16, 1873, he says: "Some fifteen years ago it was my ambition to own a copy of Bowditch's Translation of the *Mécanique Céleste*. Now I am thankful I did not have money enough to buy it. How much more elegant the original is!" Mr. Hall had a hearty contempt for the process that makes a man a senior wrangler by damming the stream of intellectual effort so as to have the beauty of a waterfall. He was the determined foe of pedantry in learning, the persistent nibbling at the shell for the sake of showing off; he struck for the kernel at the earliest possible moment. He said, "The best book from which to learn the principles of the infinitesimal calculus is Euler's *Introductio ad Analysin Infinitorum;* there is no nonsense in it." This view does not suit the modern pedagogue in mathematics, but it is characteristic of the man who uttered it. Mr. Hall indulged in the luxury of having favorite authors, and it was sometimes difficult to persuade him that somebody outside the list was worth looking at. His favorites in the line of mathematical astronomy were Lagrange, Laplace, Gauss, Hansen, and Bessel.

He had an Aristophanic vein in his composition, but only indulged it before intimate friends, and was never malicious. In the Platonic dialogue Phædrus says: "I am going for a walk outside the walls." It seemed to me that Mr. Hall was always walking outside the walls, so much did he emancipate himself from custom and tradition.

He was eminently a democratic man, and disliked the naval etiquette of putting on airs of superiority and keeping one's inferiors in rank at a distance. During his long career of work with the Clark refractor, Mr. Hall was assisted by Mr. George Anderson, who, on the score of having served in the civil war, had obtained the position of a laborer at the Observatory. Mr. Anderson was exceedingly efficient in managing the dome and the driving clock. Professor Hall threw aside all ceremony in his intercourse with Mr. Anderson. Two brothers could not have been more intimate. Professor Hall would call Mr. Anderson to the eye-piece of the telescope. "Well, George, what do you see?" George would describe. "Well, how is it situated?" George would again describe in his homely way. "Well, I think we may enter on the observing book that we both saw it."

Mr. Hall had a wholesome dread of "subjectivities." He knew that Sir William Herschel had announced four satellites of Uranus that turned out to be "subjectivities," and at Pulkova they had long observed a "subjectivity" as a companion of Procyon. He determined, if possible, to escape such mistakes.

For mental relaxation Mr. Hall went to literature. He was fond of history, frequently read novels, and also poetry to a small extent. In science he did not venture outside of mathematics, astronomy, and physics.

Mr. Hall was generous to a fault. But what a world of gratitude we owe him for his heroic, herculean perseverance! I do not suppose that he ever suffered from hunger, but he had only to look out of the window to see the wolf at the door.

Professor Hall's achievements were such that honors could not fail to be heaped upon him. Elected into the National Academy of Sciences in 1875, he served as its Home Secretary for twelve years and its Vice-President for six. He was an honorary member of nearly all the scientific societies in this country and Europe which make it a point to have honorary members. For some years he was Consulting Astronomer to the Washburn Observatory at Madison, Wisconsin. He several times delivered addresses in his capacity of president of a society, the last time being before the American Association in 1902 at Washington. He received the gold medal of the Royal Astronomical Society, the Lalande and Arago Prizes from the Academy of Sciences at Paris, the degree of LL. D. from Harvard and Yale.

The portrait at the head of this memoir represents him as he appeared in his prime. It is from a photograph taken by Mr. Peters, the photographer to the Observatory, as he was sitting at a desk in the new Observatory.

PUBLISHED WRITINGS OF ASAPH HALL,

Professor of Mathematics, U. S. Navy.

COMPILED BY WILLIAM D. HORIGAN, LIBRARIAN.

(Communicated by Captain W. J. Barnette, Superintendent of U. S. Naval Observatory.)

The following list contains all the titles that could be collected in time for the April meeting of the Academy. There are many journals, especially among those devoted solely to mathematics, in which the minor articles are not indexed. In the class of general science, some of the journals have no author-index and some have published no index whatever. To have examined these journals thoroughly would have consumed a much longer time than that allowed for the compilation of the list; therefore it may be properly inferred that the list is incomplete, though it is believed to contain every paper of importance.

It should be noted that while assistant at the Harvard College Observatory, during the years 1857 to 1862, Mr. Hall was employed on the transit circle with the passages of clock stars and with lunar culminations in connection with the Lake surveys carried on by the Engineer Corps of the Army. He observed Donati's Comet in 1858. He "thoroughly revised" the Second Series of the Harvard Zones, and with Mr. T. H. Safford prepared the Third Series for the press. The results of these labors are contained in the Annals of Harvard College Observatory, vol. 2, part 2, and in vols. 3, 4, and 6.

Excepting the appendices to the Washington Observations, comparatively few separates of Professor Hall's papers have been found. It was his custom to distribute the copies allotted to him among individuals interested in their contents, rather than to libraries likely to contain sets of the journals. Only a few of the more important reviews have been noted.

The name of the U. S. Naval Observatory has been elided from many of the titles; so it must be understood that all observations have been made at that institution unless it is otherwise stated. In entering the many papers containing equatorial observations, an attempt has been made to add to each title a list of the objects observed by Professor Hall.

Except in the mathematical formulæ or where the number of an asteroid is indicated, parentheses denote the parts of titles transposed, while brackets inclose matter inserted by the compiler. The abbreviations used for the periodicals are those of the Catalogue of Scientific Papers, compiled and published by the Royal Society of London. In the case of a few journals which have been started since the last volume of the above was published, the abbreviations have been taken from the International Catalogue of Scientific Literature.

1858.

1. Elements and ephemeris of the first comet of 1858. Astron. Journ., 5, 1856–58, p. 120.

2. Elliptic elements of Comet 1858, I. Astron. Journ., 5, 1856–58, p. 138.

3. [Elements of Comet III, 1858.] Astr. Nachr., 48, 1858, col. 331.

4. [Errata in Gauss's *Theoria motus*.] Math. Month., 1, 1858–59, pp. 115, 190.

1859.

5. [Elements of Comet 1859, I.] Astron. Journ., 6, 1859–61, p. 24.

6. Rising, setting, southing places, and eclipses of the sun and moon ; the phases and age of the moon ; the rising, setting, and southing of the most conspicuous planets and fixed stars ; equation of time ; the sun's declination ; planets' places in right ascension and declination ; the time of high water at Philadelphia, etc. Friends' Almanac for the years 1859–64 ; published by T. Ellwood Chapman, Philadelphia.

7. A number n of equal circles touch each other externally and include an area of a square feet ; to find the radii of the circles. (Prize solution of Problem III.) Math. Month., 2, 1859–60, pp. 76–77.

1860.

8. [Observations of occulations of the Pleiades, made at Cambridge, 1860, March 26.] Brünnow, Astr. Notices, No. 19, 1860, p. 147.

9. Orbit of Comet III, 1860. Brünnow, Astr. Notices, No. 24, 1861, p. 189.

10. Show that in the parabola the half sum of the radii vectores, drawn to the extremities of any arc whatever, is equal to the radius vector drawn to the summit of the diameter passing through the center of the chord and parallel to the axis, plus the part of this diameter intercepted between the arc and the chord. (Prize Problem V.) Math. Month., 3, 1860–61, p. 33.

1861.

11. Elements of Comet II, 1861 [with ephemeris]. Amer. Journ. Sci., 32, 1861, pp. 259, 260.

12. Elements of Titania [to which is added a list of typographical errors in Davis's translation of the *Theoria motus*]. Astron. Journ., 6, 1859–61, p. 191.

13. Elements of Minor Planet (66). Astron. Soc. Month. Not., 21, 1860–61, p. 240; Astr. Nachr., 55, 1861, col. 299; 57, 1862, col. 130; Brünnow, Astr. Notices, No. 25, 1861, p. 4.

14. Orbit of Comet I, 1861. Brünnow, Astr. Notices, No. 26, 1861, p. 12.

15. To change a series into a continued fraction. Math. Month., 3, 1860–61, pp. 262–268.

1862.

16. Observations upon Planet (73). [Made at Cambridge by G. P. Bond, T. H. Safford, and A. Hall.] Astr. Nachr., 57, 1862, col. 310.

17. Observations of comets and small planets made at the Observatory of Harvard College. [By G. P. Bond, T. H. Safford, and A. Hall. Mr. Hall observed comets I and II, 1861, asteroids (30), (66).] Astr. Nachr., 57, 1862, col. 353–368.

18. [Elements of the planet Clytia (73). Elements of Comet III, 1861.] Astr. Nachr., 58, 1862, col. 29–30.

19. Elements and ephemeris of Maja (66). Astr. Nachr., 58, 1862, col. 87.

20. Elements of Comet I, 1862. Astr. Nachr., 58, 1862, col. 89.

21. Ephemeris of Comet I, 1862. Astr. Nachr., 58, 1862, col. 90.

22. [Elements and ephemeris of Comet II, 1862.] Astr. Nachr., 58, 1862, col. 203-204.

23. Observations of asteroids and of the Comet II, 1862 . . . By J. Ferguson and A. Hall. [Mr. Hall observed Comet II, 1862, asteroids (32), (74).] Astr. Nachr., 59, 1862-63, col. 115-122.

24. Table of zenith distances to 80°, for the latitude of the U. S. Naval Observatory. [By James Ferguson and Asaph Hall.] Washington Obs'ns, 1861, pp. 349-358.

25. Tables of differential refraction, for the latitude of the U. S. Naval Observatory, from $+ 30°$ to $- 30°$ of declination, and to 80° of zenith distance, $d'—d$ being constant and equal to $600''$. [By James Ferguson and Asaph Hall.] Washington Obs'ns, 1861, pp. 359-374.

26. Tables of parallax for the latitude of the U. S. Naval Observatory, and a horizontal parallax of 10 seconds. [By James Ferguson and Asaph Hall.] Washington Obs'ns, 1861, pp. 434-438.

1863.

27. Observations of comets and small planets, made at the Observatory of Harvard College, Cambridge. [By G. P. Bond, T. H Safford, and A. Hall. Mr. Hall observed Comet III, 1862, Asteroid (73).] Astr. Nachr., 60, 1863, col. 33-42.

28. Observations of Comet II, 1863 . . . By James Ferguson and Asaph Hall. Astr. Nachr., 60, 1863, col. 123-126.

29. Ephemeris of Maia (66). Astr. Nachr., 60, 1863, col. 175-176.

30. Observations of asteroids . . . By James Ferguson and A. Hall. [Mr. Hall observed asteroids (48), (51), (64), (79).] Astr. Nachr., 61, 1863-64, col. 209-212.

31. Observations with the equatorial, 1862. Right ascensions and declinations of asteroids and comets, observed with the equatorial, 1862. [By James Ferguson and Asaph Hall. Commencing August 7, 1862, Mr. Hall observed Mars, Comet 1862, II, asteroids (32), (74), miscellaneous double stars.] Washington Obs'ns, 1862, pp. 405-509, 510, 581-585.

1864.

32. Observations of minor planets [and of comets] . . . By James Ferguson and Asaph Hall. [Professor Hall observed comets IV and VI, 1863, asteroids (17), (25), (34), (52), 79).] Astr. Nachr., 62, 1864, col. 311-318.

33. Observations of asteroids and of Comet II, 1864. By James Ferguson and Asaph Hall. [Professor Hall observed Comet II, 1864, asteroids (2), (3), (9), (12), (16), (36), (42).] Astr. Nachr., 63, 1864-65, col. 205-208.

1865.

34. Elements of Comet IV, 1864. Astr. Nachr., 64, 1865, col. 121-124.

35. Observations of asteroids and comets . . . [By James Ferguson and Asaph Hall. Professor Hall observed Comet IV, 1864, asteroids (1), (19 , (27), (29), (37), (38), (40), (48), (60), (64), (81).] Astr. Nachr., 64, 1865, col. 177-182.

36. Observations of asteroids and comets . . . [By James Ferguson and Asaph Hall. Professor Hall observed Faye's Comet, asteroids (5), (6), (8), (13), (14), (24), (30), (31), (32), (34), (46), (49), (56), (61), (63), (67), (72), (78), (79).] Astr. Nachr., 66, 1865-66, col. 33-42.

37. Elements and ephemeris of Terpsichore (81). Astr. Nachr., 66, 1865-66, col. 55-58.

38. On the orbit of Comet III, 1858. Astr. Nachr., 66, 1865-66, col. 137-140.

39. " From the middle point of each side of a quadrilateral in a circle a perpendicular is drawn to the opposite side, and from the middle point of each diagonal to the other diagonal. Prove that the six perpendiculars pass through one point." (Mathematical Question 2027, solved.) Lady's and Gentleman's Diary, London, 162, 1865, p. 61.

40. Solar parallax, deduced from observations of Mars with equatorial instruments. Washington Obs'ns, 1863, Appendix A, pp. LX-LXIV.

41. New elements of Nemausa. Washington Obs'ns, 1863, Appendix B, pp. LXXXIII-XC.

42. Observations with the equatorial, 1863. Right ascensions and declinations of stars, asteroids and comets observed with the equatorial, 1863. [By James Ferguson and Asaph Hall. Professor Hall observed : comets 1863, I, 1863, II, 1863, IV ; asteroids (14), (18), (39), (48), (49), (51), (55), (64), (79) ; Pleiades ; double stars ; occultations of stars by the moon.] Washington Obs'ns, 1863, pp. 217-284, 363-367.

1866.

43. Measures of the companion of Sirius . . . in 1866. [By S. Newcomb, A. Hall, J. R. Eastman, and H. P. Tuttle.] Astr. Nachr., 66, 1865-66, col. 381-382.

44. [On the determination of the solar parallax from equatorial observations of Mars.] Astr. Nachr., 68, 1866-67, col. 235-236.

45. At a given latitude (ϕ) two stars whose right ascensions and declinations are a, a', δ, δ', have the same altitude above the horizon. It is required to find the hour angle, and the conditions under which the phenomenon is possible. (Mathematical Question 2052.) Lady's and Gentleman's Diary, London, 163, 1866, p. 71 ; 164, 1867, p. 52.

46. Observations with the mural circle, 1864. [Observers : S. Newcomb, A. Hall, W. Harkness, and M. H. Doolittle.] Washington Obs'ns, 1864, pp. 105-191.

47. Observations with the equatorial, 1864. Right ascensions and declinations of stars, asteroids, and comets, observed with the equatorial, 1864. [By James Ferguson and Asaph Hall. Professor Hall observed : Comet 1863, VI ; asteroids (1), (2), (3), (9), (12), (13), (16), (17), (19), (25), (27), (29), (34), (36), (37), (40), (42), (52), (60), (79), (81) ; cluster in Præsepe.] Washington Obs'ns, 1864, pp. 203-304, 377-389.

1867.

48. Observations of certain small planets, and of the stars which have been compared with Thisbe, made with the transit circle . . . during 1866 and 1867. [Observers : S. Newcomb, A. Hall, J. A. Rogers, and C. Thirion. Professor Hall observed asteroids (13), (22), (24), (28), (34), (50), (56), (71), (84), (85), (88), (89).] Astr. Nachr., 69, 1867, col. 151-156.

49. Remarks on hyperbolic movement. Messenger of Math., 4, 1868, pp. 106-107.

50. Gauss's proof that the middle points of the three diagonals of a complete quadrilateral lie in a right line. Messenger of Math., 4, 1868, p. 137.

51. Observations with the mural circle, 1865. [Observers : S. Newcomb, A. Hall, W. Harkness, J. A. Rogers, and M. H. Doolittle.] Washington Obs'ns, 1865, pp. 149-239.

52. Observations with the prime vertical transit instrument, 1865. [Observers : S. Newcomb, A. Hall, and W. Harkness.] Washington Obs'ns, 1865, pp. 241-249.

53. Observations with the equatorial, 1865. Right ascensions and declinations of stars, asteroids, and comets observed with the equatorial, 1865. [By James Ferguson and Asaph Hall. Professor Hall observed Comet 1864 IV, Faye's Comet ; asteroids (5), (6), (8), (13), (14), (16), (24), (30), (31), (32), (34), (38), (46), (48), (49), (56), (57), (61), (63), (64), (67), (72), (78), (79), (85) ; Pleiades ; cluster in Præsepe ; occultations ; and eclipse of the sun, October 18, 1865.] Washington Obs'ns, 1865, pp. 251-347, 429-435.

1868.

54. Observations of asteroids . . . during the year 1867 [on the transit circle. Observers : S. Newcomb, A. Hall, C. Thirion, and C. Abbe. Professor Hall observed asteroids (21), (24), (25), (29), (32), (37), (41), (42), (43), (45), (47), (51), (52), (59), (60), (64), (65), (68), (71), (79), (80), (82), (84), (85), (88), (92).] Astr. Nachr., 71, 1868, col. 161-170.

55. On the positions of the fundamental stars. Astr. Nachr., 71, 1868, col. 191-192.

56. Equatorial observations . . . [Brorsen's Comet, (95) Arethusa and (98) Ianthe.] Astr. Nachr., 72, 1868, col. 45-46.

57. It is required to find the mean or average distance from the vertex of a right cone (1) to all the points in the base of the cone, (2) to all the points in the solid content of the cone. (Mathematical Question 2085.) Lady's and Gentleman's Diary, London, 165, 1868, p. 95; 166, 1869, p. 79.

58. Observations with the transit circle, 1866. [Observers: S. Newcomb, A. Hall, J. A. Rogers, and C. Thirion.] Washington Obs'ns, 1866, pp. 1-194.

59. Observations with the prime vertical transit instrument, 1866. [Observers: S. Newcomb and A. Hall.] Washington Obs'ns, 1866, pp. 269-277.

60. Observations with the equatorial, 1866. Right ascensions and declinations of stars, asteroids, and comets, observed with the equatorial, 1866. [By James Ferguson, Asaph Hall, J. R. Eastman, and some of the occultations by H. P. Tuttle. Professor Hall observed Terpsichore (81), the cluster in Præsepe, and the companion of Sirius.] Washington Obs'ns, 1866, pp. 279-338, 409-413.

1869.

61. Equatorial observations . . . [Brorsen's Comet, Encke's Comet, and asteroids (26), (38), (43), (45), (54), (59), (60), (64), (76), (80), (92), (95), (101), (102), (106).] Astr. Nachr., 74, 1869, col. 71-78.

62. "Prove that $\dfrac{2}{n} = \dfrac{2^n}{n} - 2^{n-2} + \dfrac{n-3}{2} \cdot 2^{n-4} - \dfrac{(n-4)(n-5)}{2.3} \cdot$
$2^{n-6} + \dfrac{(n-5)(n-6)(n-7)}{2.3.4} \cdot 2^{n-8}$, &c., continued to 2^1 or 2^0, according as n is odd or even." (Mathematical Question 2089, answered.) Lady's and Gentleman's Diary, London, 166, 1869, p. 84.

63. Transformations of coördinates in Hansen's method of perturbations. Messenger of Math., 5, 1871, pp. 15-23.

64. Report on observations of the total eclipse of the sun, August 7, 1869. [At station near Plover Bay, Siberia.] Washington Obs'ns, 1867, Appendix II, pp. 197-218.

1870.

65. On the secular perturbations of the planets. Amer. Journ. Sci., 50, 1870, pp. 370-372.

66. Equatorial observations . . . [of Felicitas (109)]. Astr. Nachr., 75, 1869-70, col. 321-324.

67. Supplementary notes on the observations for magnetism and position, made by the U. S. Naval Observatory Expedition to Siberia to observe the solar eclipse of August 7, 1869. Astr. Nachr., 75, 1869-70, col. 323-328; Washington Obs'ns, 1867, Appendix II, pp. 215-218.

68. New elements of Terpsichore (81). Astr. Nachr., 76, 1870, col. 123-124.

69. Equatorial observations . . . [of the asteroids (51), (61), (70), (71), (110), (111)]. Astr. Nachr., 77, 1870-71, col. 15-16.

70. "If from any point P in the hypothenuse AB of a right-angled spherical triangle ABC, perpendiculars PE, PD are drawn to CB, CA, and if the segments CE, CD be denoted by a, β respectively, prove that $\dfrac{\tan a}{\tan a} + \dfrac{\tan \beta}{\tan b} = 1$, where a, b are the sides." (Mathematical Question 2095, answered.) Lady's and Gentleman's Diary, London, 167, 1870, p. 75.

71. In Hansen's *Theory of perturbations* he makes use of what he calls the "arithmetico-geometrico mean." Thus, if A and B are two values, we have $\frac{1}{2}$ (A + B) = arith. mean = A_1; $\sqrt{AB}$ = geom. mean = B_1. Show that, by repeating this process on A_1 and B_1, and so on, the results A_n and B_n approach each other without limit. (Mathematical Question 2098, with answer.) Lady's and Gentleman's Diary, London, 167, 1870, p. 77.

72. Coördinates of a celestial body. Messenger of Math., 5, 1871, pp. 250-251.

73. Observations with the transit circle, 1867. [Observers: S. Newcomb, A Hall, J. A. Rogers, C. Thirion, and C. Abbe.] Washington Obs'ns, 1867, pp. 1-212.

74. Observations with the prime vertical transit instrument, 1867. [Observers: S. Newcomb, A. Hall, and C. Abbe.] Washington Obs'ns, 1867, pp. 303-305.

75. Catalogue of 151 stars in Præsepe. [Washington Observations, 1867] (Appendix IV). [Washington, 1870.] 38 pp., $29\frac{1}{2}$ cm. Reviewed in Messenger of Math., 5, 1871, p. 151.

1871.

76. Transit of Venus in 1874. Amer. Journ. Sci., 1, 1871, pp. 307-308.

77. On the application of photography to the determination of astronomical data. Amer. Journ. Sci., 2, 1871, pp 25-30, 154. Abstract in Washington, Phil Soc. Bull., 1, 1871-74, pp. 28-29; Smithson. Misc. Coll., 20, 1881, Art. 1, pp. 28-29.

78. On the astronomical proof of a resisting medium in space. Amer. Journ. Sci , 2, 1871, pp. 404-408. Abstract in Astron. Soc. Month. Not., 33, 1872-73, pp. 239-242.

79. Observations and elements of Comet I, 1871. Astr. Nachr., 77, 1870-71, col. 319-320.

80. Ephemeris of Terpsichore, 1869. The Washington stars of comparison for 1868. Observations of Egeria, 1864. Photographic observations of the Venus transit in 1874. Astr. Nachr., 78, 1871-72, col. 167-168.

81. "Each of three circles (radii ρ_1 ρ_2 ρ_3) touches two sides of the triangle $(a\,b\,c)$ and the nine-point circle; prove that $\dfrac{\rho_1}{r_1} + \dfrac{\rho_2}{r_2} + \dfrac{\rho_3}{r_3} = \dfrac{a^2 + b^2 + c^2}{(a + b + c)^2}$." (Mathematical Question 2113, answered.) Lady's and Gentleman's Diary, London, 168, 1871, pp. 77-78.

82. "The four sides of a quadrilateral are given; determine its form when the rectangle under the two diagonals is a maximum." (Mathematical Question 2117, answered.) Lady's and Gentleman's Diary, London, 168, 1871, p. 82.

83. Required, the integral of $\dfrac{d\,x}{\sin^2 x \cos^2 x}$. (Problem 89, with solution.) Schoolday Visitor, 15, 1871, pp. 164, 251.

84. [Introduction to the observations made with the 9.6-inch equatorial, 1868, 1869, 1870, 1871, 1873.] Washington Obs'ns, 1868, p. XL; 1869, p. XLIV; 1870, p. LXXXI; 1871, p. C; 1873, p. CI.

85. Observations with the transit circle, 1868. [Observers: S. Newcomb, A. Hall, W. Harkness, J. R. Eastman, C. Thirion, C. Abbe, and E. Frisby.] Washington Obs'ns, 1868, pp. 1-154.

86. Observations with the equatorial, 1868. [Brorsen's and Encke's comets, asteroids (26), (38), (43), (45), (54), (59), (60), (64), (76), (80), (92), (95), (98), (101), (102), (106), occultations of stars by the moon.] Washington Obs'ns, 1868, pp. 319-327.

87. Report on observations of the total solar eclipse of December 22, 1870. [Station at Syracuse, Sicily.] Washington Obs'ns, 1869, Appendix I, pp. 25-42. Quoted as to meteorological changes during totality and the appearance of the corona, in Astron. Soc. Mem., 41, 1879, pp. 213, 615, 643-644.

88. On the elements of 'the Comet I, 1871. Abstract. Washington, Phil. Soc. Bull., 1, 1871-74, p. 23; Smithson. Misc Coll., 20, 1881, Art. 1, p. 23.

89. On a curve of the fourth degree. ["Through the focus of an ellipse a right line is drawn, cutting the ellipse in the points D and E, and at the middle point of DE an indefinite right line is drawn perpendicular to DE. It is required to find the form and area of the curve that this perpendicular always touches."] Abstract. Washington, Phil. Soc. Bull., 1, 1871-74, pp. 30-31; Smithson. Misc. Coll., 20, 1881, Art. 1, pp. 30-31. See No. 112

1872.

90. Equatorial observations . . . 1871. [Comets I, II, III, and IV, 1871; asteroids (10), (12), (19), (23), (26), (29), (31), (35), (41), (47), (48), (52), (54), (55), (59), (62), (64), (68), (71), (80), (81), (88), (92), (93), (95), (109), (113), (114), (116).] Astr. Nachr., 79, 1872, col. 97-112.

91. Observations of the companion of Sirius . . . Astr. Nachr., 79, 1872, col. 247-248.

92. Observations of Planet (121) . . . Astr. Nachr., 79, 1872, col. 367-368.

93. [Letter in regard to Professor Cayley's memoir, *On the determination of the orbit of a planet from three observations*.] Astr. Nachr., 80, 1872-73, col. 13-14.

94. Elements of Comet *a*, 1871, Winnecke. Astr. Nachr., 80, 1872-73, col. 29-32.

95. On an experimental determination of π. Messenger of Math., 2, 1873, pp. 113–114. See No. 105.

96. Historical note on the method of least squares. Nature, 6, 1872, pp. 101–102, 241–242. Title only in Washington, Phil. Soc. Bull., 1, 1871–74, p. 62 ; Smithson. Misc. Coll., 20, 1881, Art. 1, p. 62.

97. Hindrances to students of mathematics. Nature, 6, 1872, p. 351.

98. [November meteors observed at Washington.] Nature, 7, 1872–73, pp. 122–123.

99. Logarithmic tables [Remarks on]. Nature, 7, 1872–73, p. 222.

100. Observations with the transit circle, 1869. [Observers : S. Newcomb, A. Hall, W. Harkness, C. Thirion, and E. Frisby.] Washington Obs'ns, 1869, pp. 1–59.

101. Observations with the equatorial, 1869. [Asteroid (109) and occultations of stars by the moon. Professor Newcomb observed one of the occultations] Washington Obs'ns, 1869, pp. 233–236.

102. Zones of stars observed . . . with the mural circle in the years 1846, 1847, 1848, and 1849. By J. H. C. Coffin, T. J. Page, Charles Steedman. [Edited and introduction written by Asaph Hall.] (Washington Observations for 1869, Appendix II.) Washington, 1872. 25 + 331 pp. 30 cm.

103. Reports on observations of Encke's Comet during its return in 1871. By Asaph Hall and William Harkness. (Washington Observations for 1870, Appendix II.) Washington, 1872. 1 pl , 49 pp. 30 cm.

104. Zones of stars observed . . . with the meridian transit instrument in the years 1846, 1847, 1848, and 1849. By Reuel Keith, Mark H. Beecher, Joseph S. Hubbard, John J, Almy, and William A. Parker. [Edited and introduction written by Asaph Hall.] (Washington Observations for 1870, Appendix IV.) Washington, 1872. 13 + 271 pp. 30 cm.

105. On the experimental determination of the ratio of the circumference to the diameter, based on the principles of the calculus of probabilities. [Title only of paper read.] Washington, Phil. Soc. Bull., 1, 1871–74, p. 62; Smithson. Misc. Coll., 20, 1881, Art. 1, p. 62. See No. 95.

1873.

106. On the determination of longitudes by moon culminations. Astron. Soc. Month. Not., 33, 1872–73, pp. 465–466.

107. Elements of (124) Alceste, and observations of (129). Astr. Nachr., 81, 1873, col. 109–110.

108. Equatorial observations . . . [April–December, 1871. Asteroids (46), (64), (80), (98), (118), (119), (121), (122), (124), (128).] Astr. Nachr., 81, 1873, col. 171–176.

109. Positions of the principal stars derived from observations made at U. S. Naval Observatory by Prof. M. Yarnall in the years 1853 to 1860. Astr. Nachr., 82, 1873, col. 43–48.

110. Observations of Comet e, 1873, Henry . . . [with elements of its orbit]. Astr. Nachr., 82, 1873, col. 243–244.

111. New elements of (124) Alceste, with opposition ephemeris, 1873. Astr. Nachr., 82, 1873, col. 261–266.

112. "EG is a focal chord in an ellipse and MP a perpendicular to it at its middle point M. Show that the curve which the line MP always touches is a three-cusped curvilinear triangle, to which the axes of the ellipse are tangents, and that its area is to that of the ellipse as $e^6 : 1 - e^2$, where e is the eccentricity." (Mathematical Question 3731, solved.) Educational Times, 25, 1872–73, (p. 272; Math. Quest. from Educational Times, 19, 1873, pp. 28–29. See also No. 89.

113. On the motion of a particle toward an attracting center at which the force is infinite. Messenger of Math., 3, 1874, pp. 144–149.

114. Observations with the transit circle, 1870. [Observers: A. Hall, W. Harkness, J. R. Eastman, E. Frisby, and O. Stone.] Washington Obs'ns, 1870, pp. 1–87.

115. Observations with the equatorial, 1870. [Asteroids (51), (61), (70), (71), (109), (110), (111); occultations of stars by the moon, two of which were observed by Professor Newcomb.] Washington Obs'ns, 1870, pp. 255–259.

116. Observations with the transit circle, 1871. [Observers: A. Hall, W. Harkness, J. R. Eastman. E. Frisby, and O. Stone.] Washington Obs'ns, 1871, pp. 1–45.

117. Observations with the equatorial, 1871. [Comets I and II, 1871, and those of Encke and Tuttle; asteroids (10), (12), (19), (23), (26), (29), (31), (35), (41), (47), (48), (52), (54), (55), (59), (62), (64), (68), (71), (80), (81), (88), (92), (93), (95), (109), (113), (114), (116); occultations of stars by the moon. Some of the occultations were observed by S. Newcomb, A. N. Skinner, and H. P. Tuttle.] Washington Obs'ns, 1871, pp. 103–116.

118. Zones of stars observed . . . with the meridian circle in the years 1847, 1848, and 1849. By James Major, Lafayette Maynard, William B. Muse. [Edited and introduction written by Asaph Hall.] (Washington Observations for 1871, Appendix I.) Washington, 1873. 9 + 162 pp. 30 cm.

1874.

119. Comets and meteors. Analyst, Des Moines, 1, 1874, pp. 17–24.

120. The chief justice of a court makes a large number of decisions. Afterward it is found that 50 per cent. of these decisions are erroneous. Required, to determine the legal knowledge of the judge. (Problem 10, with solution.) [Moral for weather prophets, fortune-tellers, etc.] Analyst, Des Moines, 1, 1874, pp. 35, 71.

121. Besselian function. Analyst, Des Moines, 1, 1874, pp. 81–84.

122. "If a, b, c, d, e, f, g, h, i, j, k be chords drawn from any point on the circumference of a circle to the eleven angles of an inscribed regular polygon of eleven sides, prove that $(a + k)(b + j)(c + i)(d + h)(e + g) = f^5$ . . . . (1)." (Problem 29, solved.) Analyst, Des Moines, 1, 1874, pp. 159–160.

123. Assuming the earth's orbit to be a circle, if a comet move in a parabola around the sun and in the plane of the earth's orbit, show that the comet cannot remain within the earth's orbit longer than 78 days (Problem 50) Analyst, Des Moines, 1, 1874, p. 212 ; 2, 1875, p. 30.

124. Equatorial observations . . . in 1873. [Comets a, b, c, d, e, and g, 1873 ; asteroids (9), (11), (21), (31), (33), (40), (43), (46), (49), (53), (58), (59), (60), (63), (67), (69), (71), (78), (81), (83), (92), (109), (112), (124), (129), (130), (131).] Astr. Nachr., 84, 1874, col. 17–28.

125. On the orbit of Alceste (124). Astr. Nachr., 84, 1874, col. 89–94.

126. [Correction to an observation of Alceste (124), December 10, 1874 ; position of comparison star for Electra (130).] Astr. Nachr., 84, 1874, col. 163–164.

127. Observations with the transit circle, 1872. [Observers: A. Hall, W. Harkness, J. R. Eastman, E. Frisby, and O. Stone.] Washington Obs'ns, 1872, pp. 1–159.

128. Observations with the equatorial, 1872. [Asteroids (46), (64), (80), (98), (118), (119), (121), (122), (124), (128) ; occultations ; companion of Sirius, which was also observed by S. Newcomb and A. N. Skinner.] Washington Obs'ns, 1872, pp. 205–211.

129. Remarks on the method adopted in writing international telegrams. [Recommended the use of Littrow's system.] Washington, Phil. Soc. Bull., 1, 1871–74, p. 101 ; Smithson. Misc. Coll., 20, 1881, Art. 1, p. 101.

1875.

130. Photographic observations of the transit of Venus. Analyst, Des Moines, 2, 1875, pp. 89–90.

131. Into how many parts can n planes divide space? (Query.) Analyst, Des Moines, 2, 1875, p. 96.

132. If the parabolic orbits of two comets intersect the circular orbit of the earth in the same two points, then if t_1 and t_2 be the times in which the comets move from one point to the other, $(t_1 + t_2)^{\frac{2}{3}} + (t_1 - t_2)^{\frac{2}{3}} = \left(\dfrac{4}{3\pi}\right)^{\frac{2}{3}}$, a year being the unit of time. (Problem 82.) Analyst, Des Moines, 2, 1875, pp. 128, 158.

133. Note on the division of space. Analyst, Des Moines, 2, 1875, p. 140.

134. Ephemeris of Terpsichore (81) for the opposition, 1876. Astr. Nachr., 86, 1875, col. 3–6.

135. Observations made with the 26-inch equatorial . . . 1875. [Observers: S. Newcomb, A. Hall, and E. S. Holden. Article signed by A. Hall, who observed Oberon, Titania, Umbriel, and the companion of Sirius.] Astr. Nachr., 86, 1875, col. 321–328.

136. On the determination of the mass of Mars. Astr. Nachr., 86, 1875, col. 327–334.

137. On the Washington observations of Flora in 1873. Astr. Nachr., 86, 1875, col. 333–336.

138. Observations of the satellites of Saturn. Astr. Nachr., 87, 1875-76, col. 177-190.

139. Observations with the 9½-inch equatorial, 1873. [Observers: A. Hall and A. N. Skinner. Professor Hall observed comets a, b, c, d, e, g, 1873; occultations; companion of Sirius; asteroids (8), (9), (21), (31), (33), (40), (43), (46), (53), (58), (59), (60), (63), (67), (69), (71), (78), (81), (83), (109), (112), (124), (129), (130), (131).] Washington Obs'ns, 1873, pp. 159-173.

140. [Report on the operations of the Transit of Venus Party at Vladivostok, 1874.] Abstract. Washington, Phil. Soc. Bull., 2, 1874-78, pp. 32-33; Smithson. Misc. Coll., 20, 1881, Art. 2, pp. 32-33. A more extended account in the New York Daily Tribune, March 26, 1875.

1876.

141. Companions of Procyon. [By Asaph Hall and Edward S. Holden. Professor Hall failed to see them.] Amer. Acad. Proc., 11, 1875-76, pp. 185-190; Astr. Nachr., 87, 1875-76, col. 241-246.

142. Approximate quadrature. Analyst, Des Moines, 3, 1876, pp. 1-10. Abstract in Washington, Phil. Soc. Bull., 2, 1874-78, pp. 48-49; Smithson. Misc. Coll., 20, 1881, Art. 2, pp. 48-49.

143. Show that the determinant $\begin{vmatrix} a & b & c & d \\ b & a & d & c \\ c & d & a & b \\ d & c & b & a \end{vmatrix}$ is divisible by $(a+b)^2 - (c+d)^2$; and by $(a-b)^2 - (c-d)^2$. (Problem 109.) Analyst, Des Moines, 3, 1876, pp. 64, 89.

144. NOTE.—["That a revolving ellipsoid of three unequal axes can be in equilibrium" was not discovered by Mr. Ivory, but by C. G. J. Jacobi, whose note is published in Poggendorff's *Annalen*, 33, 1834, p. 229.] Analyst, Des Moines, 3, 1876, p. 127.

145. In a spherical triangle are given the sum of each angle and the side opposite, to solve the triangle. (Problem 121.) Analyst, Des Moines, 3, 1876, pp. 132, 157.

146. Integrate the expression $\dfrac{x \cdot dx}{(x^3 + 8)\sqrt{(x^3 - 1)}}$. (Problem 132) [with note on Legendre's and Clausen's methods of solution]. Analyst, Des Moines, 3, 1876, pp. 163, 191; 4, 1877, p. 55.

147. Observations of the satellites of Neptune and Uranus [and the companion of Sirius]. Astr. Nachr., 88, 1876, col. 131-138.

148. Equatorial observations . . . [End of the eclipse of the sun, September 28, 1875. Occultations of Saturn, August 6 and September 2, 1876. By A. Hall, W. Harkness, E. S. Holden, and E. Frisby.] Astr. Nachr., 88, 1876, col. 297-298.

1877.

149. Elements of Hyperion, with ephemeris for 1877. Astr. Nachr., 90, 1877, col. 7-12.

150. Observations of the satellites of Saturn [June–December, 1876]. Astr. Nachr., 90, 1877, col. 129–138.

151. On the rotation of Saturn. Analyst, Des Moines, 4, 1877, pp. 36–42; Astr. Nachr., 90, 1877, col. 145–150. Synopsis in Astron. Soc. Month. Not., 38, 1877–78, pp. 209–210. Reviewed in Nature, 16, 1877, pp. 363–364.

152. The approximate value of the definite integral $\int_0^{\frac{\pi}{2}} \sqrt{\sin \phi} \, . \, d\phi$ is 1.198. Is there a convenient way of computing this numerical value? Query. Analyst, Des Moines, 4, 1877, p. 48. See No. 163.

153. P and Q being functions of x, find the conditions that the equation $y \, dy + (P - Qy) \, dx = 0$, is made integrable by the factor $\dfrac{y}{[y + f(x)]^n}$ and determine the form of $f(x)$. (Problem 160.) Analyst, Des Moines, 4, 1877, p. 64.

154. Differential equations of Problem 165. [The motion of two planets.] Analyst, Des Moines, 4, 1877, pp. 146–147.

155. Discovery of the satellites of Mars. [Letter to Mr. Glaisher.] Astron. Soc. Month. Not., 38, 1877–78, pp. 205–208.

156. On the appearance of Saturn's rings. Astr. Nachr., 90, 1877, col. 151–154. Abstract in Washington, Phil. Soc. Bull., 2, 1874–78, p. 94; Smithson. Misc. Coll., 20, 1881, Art. 2, p. 94.

157. Observations made with the 26-inch refractor . . . [Satellites of Neptune and Uranus, the companion of Sirius; by A. Hall and E. S. Holden. Double stars, by A. Hall.] Astr. Nachr., 90, 1877, col. 161–166.

158. Shadow of a planet. Astr. Nachr., 90, 1877, col. 305–314.

159. [Correction to the published observation of the satellites of Mars, made August 20, 1877; and on the effect of observing with a large refractor on estimates of magnitudes.] Astr. Nachr., 90, 1877, col. 361–362.

160. Observations of the satellites of Mars [August 11 to September 16, 1877]. Astr. Nachr., 91, 1877–78, col. 11–14.

161. On the position of the south-polar spot of Mars. Astr. Nachr., 91, 1877–78, col. 219–224.

162. [On the stars in the halo of light surrounding Sirius.] Astr. Nachr., 91, 1877–78, col. 223–224.

163. Find the approximate value of the integral $\int_0^{\frac{1}{2}\pi} (\sin \theta)^{\frac{1}{2}} \, d\theta$. (Mathematical Question 5249.) Educational Times, 30, 1877–78, pp. 21, 91; Math. Quest. from Educational Times, 28, 1877, p. 19. See No. 152.

164. [Condition of the atmosphere in regard to astronomical observations at the Old Naval Observatory. Letter to the Superintendent, dated September 8, 1877.] In Reports on the removal of the United States Naval Observatory; Rear-Admiral John Rodgers. Washington, 1877, p. 9; also in Report Sec. Navy, 1877, p. 313.

165. [Letter to the Superintendent in relation to change of organization of the Naval Observatory, dated November 24, 1877. Signed by M. Yarnall, A. Hall, and others.] In Reports on the removal of the United States

Naval Observatory, Rear Admiral John Rodgers. Washington, 1877, pp. 17–18; Report Sec. Navy, 1877, pp. 321–322.

166. [Results of measures of Sirius and companion, March 25 and 26, 1874, with the 26-inch equatorial.] Washington Obs'ns, 1874, p. 290.

167. Bright spot which had recently become visible on the ball of Saturn. Abstract. Washington, Phil. Soc. Bull., 2, 1874–78, p. 102; Smithson. Misc. Coll., 20, 1881, Art. 2, p. 102.

1878.

168. [Circular letter, signed by M. Yarnall, A. Hall, and others, addressed to the members of the National Academy of Sciences, and taking exception to certain reflections upon the professional astronomers of the Naval Observatory, contained in a circular previously received by members of the Academy.] Washington, D. C., January 24, 1878, 1 p., 25½ cm.

169. Report of the committee [of the National Academy of Sciences] on proposed changes in the Nautical Almanac. [Signed by J. E. Hilgard, J. H. C. Coffin, A. Hall, and others.] American Ephemeris and Nautical Almanac, 1882, pp. 518–519.

170. Observations of the brightness of the satellites of Uranus. [By A. Hall and E. S. Holden.] Amer. Journ. Sci., 15, 1878, pp. 195–197.

171. Center of gravity of the apparent disk of a planet. Analyst, Des Moines, 5, 1878, pp. 44–45; Astron. Soc. Month. Not., 38, 1877–78, pp. 122–123.

172. Given $x = a \cos A \pm r_1$, $y = a \sin A \pm r_2$, where r_1 and r_2 are the probable errors of x and y. Required the probable errors of a and A. (Problem 201.) Analyst, Des Moines, 5, 1878, pp. 64, 92.

173. Observations of stars around the ring-nebula in Lyra. Astr. Nachr., 92, 1878, col. 27–28.

174. Names [and approximate elements] of the satellites of Mars. Astr. Nachr., 92, 1878, col. 47–48.

175. Corrections to observations of comets. [Tuttle's Comet and Comet c, 1873, Borrelly.] Astr. Nachr., 92, 1878, col. 365–366.

176. Observations with the 26-inch refractor . . . [Satellites of Saturn; disappearance of the ring; satellites of Uranus; Venus; transit of Mercury, May 6, 1878; companion of Sirius. Of these the observations of Oberon and Titania, two observations of Mimas, and six of the companion of Sirius were made by Professor Holden.] Astr. Nachr., 93, 1878, col. 65–70.

177. Mathematical Question 5522. [Statement not accessible.] Educational Times, 31, 1878–79, p. 21.

178. Mathematical Question 5596. [Statement not accessible.] Educational Times, 31, 1878–79, p. 113.

179. [Introduction to the observations made with the 26-inch equatorial, 1875–1890.] Washington Obs'ns, 1875, pp. LXXXIII–LXXXIV; 1876, pp. XCI–XCIII; 1877, p. LXXXV; 1878, p. LXXI; 1879, p. LXXI; 1880, p. LXXI; 1881, p. LXXV; 1882, p. LXIII; 1883, p. LXXIX;

1884, p. XCIII; 1885, p. CIII; 1886, p. XCVII; 1887, p. LXXXVII 1888, p. A91; 1889, p. LXVII; 1890, p. LI.

180. Observations made with the 26-inch equatorial, 1875. [Observers: S. Newcomb, A. Hall, E. S. Holden, C. L. Doolittle, D. P. Todd, and H. P. Tuttle. Professor Hall observed the satellites of Saturn, of Uranus, and of Neptune; diameter of Jupiter, diameters of the rings of Saturn, companion of Sirius, double stars, and nebulæ.] Washington Obs'ns, 1875, pp. 283-366.

181. Observations of the solar eclipse September 28, 1875 [made with the 9.6-inch equatorial. A. Hall, observer; H. M. Paul, recorder]. Washington Obs'ns, 1875, p. 372.

182. Observations and orbits of the satellites of Mars; with data for ephemerides in 1879. Washington, 1878. 46 pp. 29½ cm. A few copies of this paper were bound in with the Washington Observations for 1875; the others were issued separately. There is an abridged French translation by Paul Guieysse, in Liouville, Journ. Math., 5, 1879, pp. 143-162. The discovery of the satellites of Mars was officially announced in the "Letter to the Hon. R. W. Thompson, Secretary of the Navy . . . [signed by] John Rodgers, Rear-Admiral and Superintendent." (Washington, August 21, 1877.) 3 pp., 25½ cm. Reprinted in the Amer. Journ. Sci., 14, 1877, pp. 326-327; Astron. Soc. Month. Not., 37, 1876-77, pp. 443-445; Astr. Nachr., 90, 1877, col. 273-276. A copy of the dispatch sent out by the Smithsonian Institution is contained in the above letter; a German translation, in the Astr. Nachr., 90, 1877, col. 189-190, with later data in English, col. 239-240; a French translation in Paris, Acad. Sci. Compt. Rend., 85, 1877, p. 437.

183. Results of his search for satellites of Mars. Abstract. Washington, Phil. Soc. Bull., 2, 1874-78, p. 186; Smithson. Misc. Coll., 20, 1881, Art. 2, p. 186.

184. [Remarks on planetary motions.] Washington, Phil. Soc. Bull., 2, 1874-78, pp. 188, 189, 192; Smithson. Misc. Coll., 20, 1881, Art. 2, pp. 188, 189, 192.

185. [Remarks on the transit of Mercury, 1878.] Washington, Phil. Soc. Bull., 2, 1874-78, p. 199; Smithson. Misc. Coll., 20, 1881, Art. 2, p. 199.

186. On the supposed discovery of a trans-Neptunian planet at the U. S. Naval Observatory in 1850. Washington, Phil. Soc. Bull., 3, 1878-80, pp. 20-21; Smithson. Misc. Coll., 20, 1881, Art. 3, pp. 20-21.

187. Report on the orbits of the satellites of Mars. [Title only.] Washington, Proc. Nation. Acad., 1, 1863-94, p. 129; Report, 1878-79, p. 2.

1879.

188. On the observations of double stars. Amer. Assoc. Proc., 28, 1879, pp. 65-73.

189. If in the plane xy the directions of the forces P and P¹ make with the axis of x the angles a and a^1, and the direction of their resultant the angle A, and if we denote by p, p^1, and r the perpendiculars from the origin on

these directions, we have the relation $r \sin (a^1 - a) + p \sin (A - a^1) + p^1 \sin (a - A) = 0$. (Problem 248, with solution.) Analyst, Des Moines, 6, 1879, pp. 31, 62.

190. Stellar parallax. Analyst, Des Moines, 6, 1879, pp. 33–40.

191. When we descend below the surface of the earth, does the earth's attractive force increase or diminish? (Query 1.) Analyst, Des Moines, 6, 1879, pp. 64, 85.

192. "If the given quantities x_1, x_2, x_3, x_4 have the probable errors r_1, r_2, r_3, r_4, respectively, find the probable error r of the quantity x when $x_1 : x_2 : : x_3 + x : x_4 + x$." (Problem 256, solved.) Analyst, Des Moines, 6, 1879, p. 93.

193. Find the moments of inertia of an elliptic disk about a straight line in the plane of the disk and parallel to (1) the axis of x, (2) the axis of y, the equation of the disk being $ax^2 + 2\,bxy + cy^2 + 2\,dx + 2\,ey + f = 0$. (Problem 275, with solution.) Analyst, Des Moines, 6, 1879, pp. 128, 158–159; also proposed as Mathematical Question 6041, in the Educational Times, 32, 1879, pp. 243, 268; Math. Quest. from Educational Times, 32, 1879, p. 50.

194. Motion of a satellite. Analyst, Des Moines, 6, 1879, pp. 129–139.

195. On a theorem of Lambert's. Analyst, Des Moines, 6, 1879, pp. 171–173.

196. "If from any point in the plane of a parallelogram perpendiculars are let fall on the diagonal and on the two sides that contain this diagonal, the product of the diagonal by its perpendicular is equal to the sum of the products of the sides by their respective perpendiculars if the point falls outside of the parallelogram, or to their difference if it lies within the parallelogram." Varignon's Theorem: *Mécanique analytique*, Tome 1, p. 13. (Problem 283, selected by Professor Hall.) Analyst, Des Moines, 6, 1879, p. 190; Lagrange quoted, 7, 1880, p. 27.

197. Extracts from a letter to the Astronomer Royal. [On the elements of some of the planets.] Astron. Soc. Month. Not., 39, 1878–79, pp. 373-374.

198. Note on Hyperion. Astron. Soc. Month. Not., 39, 1878–79, p. 517.

199. Observations of Hyperion. Astr. Nachr., 94, 1878–79, col. 221-222.

200. [Letter to the editor stating that an observation of the companion of Sirius made at the Naval Observatory, 1873, November 29, which gives $p = 59°4\ s = 12''27$ had been erroneously ascribed to the writer, who wished to say, therefore, that the observation was not made by him.] Astr. Nachr., 94, 1878–79, col. 383-384.

201. Motion of Hyperion. Astr. Nachr., 95, 1879, col. 109–112.

202. Note on Saturn. Astr. Nachr., 95, 1879, col. 191–192.

203. Observations of the companion of Sirius. [By A. Hall and E. S. Holden.] Astr. Nachr., 95, 1879, col. 329–330.

204. Mathematical Question 5958. [Statement not accessible.] Educational Times, 32, 1879, p. 152.

205. Mathematical Question 6068. [Statement not accessible.] Educational Times, 32, 1879, p. 269.

206. Mathematical Question 6119. Educational Times, 32, 1879, p. 316. Same as No. 221.

207. Trans-Neptunian planet. Nature, 19, 1878-79, p. 481.

208. Les satellites de Mars en 1879. Paris, Acad. Sci. Compt. Rend., 89, 1879, pp. 776-778.

209. Report on telescopic observations of the transit of Mercury, May 5-6, 1878. Washington Obs'ns, 1876, Appendix II, pp. 3-7.

210. On the satellites of Saturn. [Hyperion and Titan.] Abstract. Washington, Phil. Soc. Bull., 3, 1878-80, p. 26; Smithson. Misc. Coll., 20, 1881, Art. 3, p. 26.

211. Notes on the orbits of Titan and Hyperion. Abstract. Washington, Phil. Soc. Bull., 3, 1878-80, pp. 40-41; Smithson. Misc. Col., 20, 1881, Art. 3, pp. 40-41.

212. Satellites of Mars in 1879. [Title only.] Washington, Proc. Nation. Acad., 1, 1863-94, p. 166; Report, 1879-80, p. 6. See Nos. 208, 218.

1880.

213. [On the progress of astronomy.] Address, Vice-President Section A. Amer. Assoc. Proc., 29, 1880, pp. 99-114; Nature, 22, 1880, pp. 570-574; Observatory, London, 3, 1879-80, pp. 594-601, 640-645; Science, Michels, 1, 1880, pp. 123-127. French translation in Les Mondes, 54, 1881, pp. 26-33.

214. Note on the companion of Sirius. Amer. Journ. Sci., 19, 1880, pp. 457-458.

215. Given, the common astronomical equations: $\tan (\lambda - \Omega) = \cos i \tan u$, $\sin \beta = \sin i \sin u$, eliminate u, and show in this manner that $\tan \beta = \tan i \sin (\lambda - \Omega)$. (Problem 295, with solution.) Analyst, Des Moines, 7, 1880, pp. 31, 62.

216. Parallel chords in an ellipse. Analyst, Des Moines, 7, 1880, pp. 82-83.

217. (1.) If v be the potential function, we have the equation given by Laplace, $\dfrac{d^2v}{dx^2} + \dfrac{d^2v}{dy^2} + \dfrac{d^2v}{dz^2} = 0$, which holds for a point outside the attracting body. For a point inside this body we have the equation given by Poisson, $\dfrac{d^2v}{dx^2} + \dfrac{d^2v}{dy^2} + \dfrac{d^2v}{dz^2} = -4\pi\rho$. What is the value of the right-hand side of this equation for a point on the surface of the attracting body? Moigno says, "In this case the expression will have in reality eight distinct values." Statics, p. 460.

(2.) Given a hemispherical dome turning about a pin at the top, and having a slit extending from the horizon to the zenith, can a telescope be placed inside this dome in such a position that every point of the heavens can be seen through the telescope? (Query.) Analyst, Des Moines, 7, 1880, pp. 135, 161-162.

218. Observations of the satellites of Mars. Astron. Soc. Month. Not., 40, 1879-80, pp. 272-283.

219. Note on β Leporis. Astr. Nachr., 96, 1879–80, col. 239–240.

220. Observations of the companion of Sirius. Astr. Nachr., 97, 1880, col. 319–320.

221. If Δ_1, Δ_2, Δ_3, Δ_4 be the lengths of four parallel chords in an ellipse, and if (1.2), (1.3), (1.4), (2.3), (2.4), (3.4) denote the distances between these chords, prove that $+ (2.3)\ (2.4) \cdot (3.4) \cdot \Delta_1{}^2 - (1.3)\ (1.4)\ (3.4)\ \Delta_2{}^2 + (1.2)\ (1.4) \cdot (2.4) \cdot \Delta_3{}^2 - (1.2)\ (1.3)\ (2.3)\ \Delta_4{}^2 = 0$. (Mathematical Question 6119, selected from Savary in the *Conn. des Temps*, 1830, Add. p. 65.) Educational Times, 33, 1880, p. 21. Same as 206.

222. "If two bodies revolve about a center, acted upon by a force proportional to the distance from the center, and independent of the mass of the attracted body, prove that each will appear to the other to move in a plane, whatever be the mutual attraction." (Mathematical Question 5968, solved.) Educational Times, 33, 1880, p. 309.

223. Notes of observations of contact [made at Vladivostok]. In observations of the transit of Venus, December 8–9, 1874, made and reduced under the direction of the Commission created by Congress; edited by Simon Newcomb. Part I. Washington, 1880, pp. 145–146.

224. Comets [visible in October, 1880]. Science, Michels, 1, 1880, p. 214.

225. Comet e, 1880. Science, Michels, 1, 1880, p. 259.

226. Tycho Brahe's new star. Science, Michels, 1, 1880, pp. 274–275.

227. Swift's Comet. Science, Michels, 1, 1880, p. 283.

228. Illustration. [In regard to the controversy between Professor Tait and Mr. Herbert Spencer.] Science, Michels, 1, 1880, pp. 309–310.

229. Tempel-Swift Comet. Science, Michels, 1, 1880, p. 330.

230. Observations made with the 26-inch equatorial, 1876. [Observers: A. Hall, E. S. Holden, and others. Professor Hall discovered and observed the white spot on Saturn, observed the satellites of Saturn, of Uranus, and of Neptune, occultations of Saturn, the companion of Sirius, double stars, &c.] Washington Obs'ns, 1876, pp. 309–400.

231. Report on the total solar eclipse of July 29, 1878. [As chief of party at La Junta, Colorado.] Washington Obs'ns, 1876, Appendix III, pp. 251–257.

1881.

232. A comet moves about the sun in a given parabolic orbit. Find the right ascension and declination of the point on the heavens towards which the comet approaches as it recedes from the sun and the earth. (Problem 338.) Analyst, Des Moines, 8, 1881, p. 31; Question 6660, Educational Times, 34, 1881, p. 123.

233. Notes on Gauss' *Theoria motus*. Analyst, Des Moines, 8, 1881, pp. 83–88.

234. Given $z = a \sin (x + \alpha) + b \sin (y + \beta)$, reduce z to the form $z = D \sin \frac{1}{2} (x + \alpha + y + \beta + \delta)$. (Problem 347.) Analyst, Des Moines, 8, 1881, pp. 103, 130; Question 6740, Educational Times, 34, 1881, p. 171; 37, 1884, p. 329.

235. "Observations on the motions of the sun-spots have also established the fact that the sun is not a fixed body, around which the earth revolves, but that it has a motion of its own through space." *Physiography*, by T. H. Huxley, 2d ed., p. 365. How can the above fact be determined by observations of the sun-spots? (Query.) Analyst, Des Moines, 8, 1881, p. 104; Science, Michels, 2, 1881, p. 215.

236. "Given the angles A, B and C of a plane triangle, and $d \log a$, $d \log b$, and $d \log c$; a, b, c being the sides respectively. What are the corresponding values dA, dB, and dC expressed in seconds of arc?" (Problem 358, solved.) Analyst, Des Moines, 8, 1881, p. 165.

237. Secular displacement of the orbit of a satellite. Analyst, Des Moines, 8, 1881, pp. 177–187.

238. Centrifugal tides. Analyst, Des Moines, 8, 1881, pp. 188–189.

239. [Obituary notice of Dr. Carl Rudolf Powalky, 1817–1881.] Astr. Nachr., 100, 1881, col. 159–160.

240. Observations of comets. [Faye's Comet and Comet e, 1880.] Astr. Nachr., 100, 1881, col. 273–278.

241. Data for ephemerides of the satellites of Mars in the opposition of 1881. Astr. Nachr., 100, 1881, col. 277–280; Science, Michels, 2, 1881, p. 543.

242. Observations of Hyperion [with ephemeris]. Astr. Nachr., 100, 1881, col. 279–282, 351.

243. Satellites of Mars in 1881. Astr. Nachr., 101, 1881–82, col. 121–122.

244. Sobre la densidad de Saturno. Cronica Científica, Barcelona, 4, 1881, p. 90.

245. Notes on double stars [e and α Lyræ]. Observatory, London, 4, 1881, pp. 281–282.

246. Lunar eclipse, June 11, 1881. [Observations.] Observatory, London, 4, 1881, p. 282.

247. Brightness of the satellites of Mars. Observatory, London, 4, 1881, p. 361.

248. [Partial solar eclipse of December 31, 1880. Observation of last contact.] Science, Michels, 2, 1881, p. 8.

249. Intra-Mercurial planets. Science, Michels, 2, 1881, pp. 202–203.

250. Satellites of Mars. [Observation made November 20, 1881.] Science, Michels, 2, 1881, p. 557.

251. Observations made with the 26-inch equatorial, 1877. [Observers: A. Hall and E. S. Holden. Professor Hall discovered and observed the satellites of Mars; observed the satellites of Saturn and of Neptune, Saturn's ring, white spot on the ball of Saturn, companion of Sirius, double stars, Ring nebula, &c.] Washington Obs'ns, 1877, pp. 183–231.

252. Observations of double stars . . . [Part I, 1863, 1875–79. Washington Observations, 1877, Appendix VI.] Washington, 1881. 144 pp., 29½ cm. Reviewed by H. A. Newton, Amer. Journ. Sci., 22, 1881, pp. 84–85; by E. B. Knobel, Astron. Soc. Month. Not., 42, 1881–82, pp. 179–180; Nature, 25, 1881–82, p. 122.

1882.

253. [Resolutions upon the death of Rear Admiral John Rodgers, adopted by the gentlemen attached to the U. S. Naval Observatory and signed by W. T. Sampson, Asaph Hall, and others.] Washington, May 10, 1882. 1 p., engraved, 33 cm.

254. Parallax of α Lyræ and 61 Cygni. Amer. Assoc. Proc., 31, 1882, pp. 93–99; Sidereal Messenger, 2, 1883–84, pp. 1–8.

255. "Show that $\int_0^{\frac{1}{2}\pi} \frac{\sqrt{(1-c)}.\,d\theta}{1-c\cos^n\theta} = \frac{\pi}{\sqrt{(2n)}}$ when c is indefinitely nearly equal to unity, n being a positive quantity." (Problem 365, solved.) Analyst, Des Moines, 9, 1882, p. 26.

256. "Show that $\int_0^a dx \int_0^x \phi(x, y).\,dy = \int_0^a dy \int_y^a \phi(x, y)\,dx$."— Dirichlet's theorem. (Problem 384, selected.) Analyst, Des Moines, 9, 1882, pp. 32, 62.

257. Note on Problem 374.—["Prove, 1st, that the probable value of any tabular value in a table of logarithms, trigonometric functions, etc., is 0.25 of a unit of the last decimal place, supposing this place correct to the nearest unit; 2d, that the average of the squares of probable errors of interpolated values depending on first differences only is $\frac{1}{3}$ $(0.25)^2$."] Analyst, Des Moines, 9, 1882, p. 48.

258. Correction of Barlow's *Tables* . . . De Morgan's edition, London, 1875. Analyst, Des Moines, 9, 1882, p. 64.

259. Given $\log 91 = 1.95904 \pm r$, $\log 92 = 1.96379 \pm r$, find $\log 91.5$ to five decimals, by simple proportion from the difference, and find the probable error of this logarithm. (Problem 391, with solution.) Analyst, Des Moines, 9, 1882, pp. 64, 94; also proposed as Problem 7030, Educational Times, 35, 1882, p. 129.

260. In a plane passing through the center of the sun, 12 right lines are drawn from this center, making an angle of 30° with each other. On each of these lines three homogeneous spherical bodies are placed at distances respectively of 10, 20, and 30 from the center of the sun, the distance from the earth to the sun being the unit of distance. The mass of each of these bodies being equal to that of the sun, what will be the velocity of a particle that starts from an infinite distance and moves in a right line towards the center of the sun, and perpendicularly to the plane of the bodies, when the particle is at a distance of 0.01 from the center of the sun, the law of attraction being that of Newton? (Problem 400, with solution.) Analyst, Des Moines, 9, 1882, pp. 96, 126.

261. "Given $\phi(x^2)\,\phi(y^2) = \phi(x'^2)\,\phi(y'^2)$ and $x^2 + y^2 = x'^2 + y'^2$, to determine the form of the function denoted by ϕ." (Problem 397, solved.) Analyst, Des Moines, 9, 1882, pp. 123–124.

262. Density of the earth. Analyst, Des Moines, 9, 1882, pp. 129–132.

263. "A smooth horizontal disk revolves with the angular velocity $\sqrt{\mu}$ about a vertical axis, at which is placed a material particle attracted to a certain point of the disk by a force whose acceleration is $\mu \times$ distance;

prove that the path on the disk will be a cycloid.—Routh's *Rigid dynamics*, p. 163." (Problem 396, solved.) Analyst, Des Moines, 9, 1882, p. 153.

264. Transform the definite integral $\int_b^a \phi\,(x)\,.\,dx$, so that the limits of integration shall be m and n. (Problem 420.) Analyst, Des Moines, 9, 1882, p. 195; 10, 1883, p. 29.

265. Conjunctions of the interior satellites of Saturn. Astron. Soc. Month. Not., 42, 1881-82, p. 308.

266. Observations of the companion of Sirius . . . [Observers : A. Hall and E. Frisby.] Astron. Soc. Month. Not., 42. 1881-82, pp. 323-324.

267. Note on meteoric astronomy. Astr. Nachr., 101, 1881-82, col. 351-352.

268. Note on double stars. Astr. Nachr., 102, 1882, col. 91-92.

269. The Greenwich observations of γ Draconis, made with the reflex zenith tube. Astr. Nachr., 102, 1882, col. 143-144.

270. Observations of the satellites of Mars in the opposition of 1881. Astr. Nachr., 102, 1882, col. 217-220.

271. Superior conjunctions of Hyperion, 1882. Astr. Nachr., 102, 1882, col. 383-384.

272. Note on ϵ Lyræ. Observatory, London, 5, 1882, p. 290.

273. Sur l'orbite de Japhet. Paris, Acad. Sci. Compt. Rend., 95, 1882, pp. 168-171.

274. Note on o^2 Eridani. Sidereal Messenger, 1, 1882-83, p. 94.

275. Observations with the 26-inch equatorial, 1878. [Observers: A. Hall and E. S. Holden. Professor Hall observed Umbriel, satellites of Saturn, Saturn's rings, double stars, &c.]—Results of observations made with the 26-inch equatorial, 1878. Washington Obs'ns, 1878, pp. 63-98.

276. Parallax of α Lyræ and 61 Cygni. [Washington Observations for 1879, Appendix I.] Washington, 1882. 64 pp., 29½ cm. Reviewed in L'Astronomie, 4, 1885, p. 311; Bul. Astr., Paris, 1, 1884, pp. 198-199; Copernicus, 3, 1883-84, pp. 188-189; by R. S. Ball, Observatory, 6, 1883, pp. 60-61.

1883.

277. Kepler's problem. Analyst, Des Moines, 10, 1883, pp. 65-66. Reviewed in Sidereal Messenger, 2, 1883-84, p. 132.

278. "A lamina is bounded on two sides by two similar ellipses, the ratio of the axes in each being m, and on the other two sides by two similar hyperbolas, the ratio of the axes in each being n. These four curves have their principal diameters along the co-ordinate axes. Prove that the product of inertia about the co-ordinate axes is $\dfrac{(a^2 - a'^2)\,(\beta^2 - \beta'^2)}{4\,(m^2 + n^2)}$ where aa', $\beta\beta'$ are the semi-major axes of the curves.—Routh's *Rigid dynamics*." (Problem 440, solved.) Analyst, Des Moines, 10, 1883, p. 127.

279. "Given five equations, $x_1{}^2 + x_2{}^2 + x_3{}^2 = 3\beta^2$, $y_1{}^2 + y_2{}^2 + y_3{}^2 = 3a^2$, $x_1y_1 + x_2y_2 + x_3y_3 = 0$, $x_1 + x_2 + x_3 = 0$, $y_1 + y_2 + y_3 = 0$. Eliminate x_2y_2 x_3y_3, and show that $a^2x_1{}^2 + \beta^2y_1{}^2 = 2 a^2\beta^2$.—Routh's *Dynamics*, 4th edition, Article 38." (Problem 444, solved.) Analyst, Des Moines, 10, 1883, p. 157.

280. Observations . . . [Conjunctions of the satellites of Saturn, the companion of Sirius, of which eight of the former and seven of the latter were made by Professor Frisby.—Note on the Great Comet, *b*, 1882.] Astron. Soc. Month. Not., 43, 1882–83, pp. 330–331.

281. Note on the mass of Saturn. Astron. Soc. Month. Not., 44, 1883–84, pp. 6–8. Synopsis in Astron. Reg., 21, 1883, pp. 276–277; noticed in Nature, 29, 1883–84, p. 185.

282. [Error in temperature coefficient for the micrometer screw of the Washington 26-inch refractor, which was applied in reducing the observations for the parallax of *a* Lyræ and 61 Cygni.] Astr. Nachr , 105, 1883, col. 271–272.

283. Observations of the companion of Sirius. [February and March, 1883. By A. Hall and E. Frisby.] Sidereal Messenger, 2, 1883–84, p. 127.

284. Constant of aberration and the solar parallax. Sidereal Messenger, 2, 1883–84, pp. 165–169.

285. Observations made with the 26-inch equatorial, 1879. [Observers : A. Hall and E. S. Holden. Professor Hall observed Ariel, Titan, Hyperion, Iapetus, satellites of Mars, double stars.]—Results of observations with the 26-inch equatorial, 1879. Washington Obs'ns, 1879, pp. 83–142.

286. Observations of the Great Comet of 1882. Washington Obs'ns, 1880, Appendix I, pp. 23–24, 29.

287. [On the science of mathematics and the "advantages of putting a question in a mathematical form."] Inaugural address of the Chairman of the Mathematical Section. Washington, Phil. Soc. Bull., 6, 1883, pp. 117–119; Smithson. Misc. Coll., 33, 1888, Art. 1, pp. 117–119.

288. [Criteria which have been proposed for the rejection of doubtful observations.] Abstract. Washington, Phil. Soc. Bull., 6, 1883, pp. 155–156; Smithson. Misc. Coll., 33, 1888, Art. 1, pp. 155-156.

1884.

289. Determination of the mass of a planet from the relative position of two satellites. Ann. Math., 1, 1884–85, pp. 1–4. Abstract in Washington, Phil. Soc. Bull., 6, 1883, pp. 132–133; Smithson. Misc. Coll., 33, 1888, Art. 1, pp. 132–133.

290. A horizontal wind blows against a hemispherical dome of radius R'. The pressure exerted by the wind on a plane surface normal to its direction is P pounds to the square foot; on a surface oblique to its direction the pressure exerted is normal, but it is reduced in the ratio $1 : 1 + \frac{1}{4} \tan^2 i$, where *i* is the angle of incidence. (Poncelet, *Mécanique industrielle*, 403.) It is required to find the magnitudes and the points of application of the horizontal and vertical components of the resultant wind-pressure. (Exercise 1, solved.) Ann. Math., 1, 1884–85, pp. 44–47.

291. The result $- \dfrac{p^2 q^2 + 4 p^2 r - 8 q^3 + 2 pqr + 9 r^2}{(r - pq)^2}$ is given as the equivalent of the function $\left(\dfrac{\beta - \gamma}{\beta + \gamma} \right)^2 + \left(\dfrac{\gamma - \alpha}{\gamma + \alpha} \right)^2 + \left(\dfrac{\alpha - \beta}{\alpha + \beta} \right)^2$, where α, β, γ are roots of the cubic $x^3 + px^2 + qx + r = 0$. Is this result correct? (Exercise 12.) Ann. Math., 1, 1884–85, pp. 48, 88, 112–113.

292. In his work *Die lineale Ausdehnungslehre, ein neuer zweig der Mathematik*, p. 65, Grassmann says: "Lagrange führt in seiner *Méc. anal.*, p. 14 der neuen Ausgabe, einen Satz von Varignon an, dessen er sich zur Verknüpfung der verschiedenen Principien der Statik bedient. * * * * Dieser Satz ist, wie sich sogleich zeigen wird, unrichtig." In what way is this theorem incorrect as used by Todhunter and others? (Exercise 17.) Ann. Math., 1, 1884–85, pp. 71, 92.

293. Observations . . . [Conjunctions of the satellites of Saturn, satellites of Mars, and companion of Sirius.] Astron. Soc. Month. Not., 44, 1883–84, pp. 358–361.

294. Motion of Hyperion. Astron. Soc. Month. Not., 44, 1883–84, pp. 361–365. Abstract in Science, 4, 1884, pp. 155–156.

295. Uniform ephemeris of the clock stars. Astr. Nachr., 109, 1884, col. 145–146; Observatory, London, 7, 1884, pp. 220–221. Abstract in Sidereal Messenger, 3, 1884, p. 180.

296. Note on the latitude of the Naval Observatory. Astr. Nachr., 110, 1884–85, col. 129–132.

297. Lettre. [Les éphémérides de Mimas et Hypérion par M. Marth.—La limite supérieure de la masse de Titan par M. Tisserand.] Bul. Astr., Paris, 1, 1884, p. 478.

298. If a planet be spherical and ϕ be the angle at the planet between the earth and the sun and a the radius of the sphere, prove that the distance of the centroid of the planet's apparent disk from its true center will be $\dfrac{8\,a}{3\,\pi} \sin^2 \tfrac{1}{2}\, \phi$ when the planet is gibbous, $\dfrac{8\,a}{3\,\pi} \cos^2 \tfrac{1}{2}\, \phi$ when the planet is crescent. (Mathematical Question 5522.) Educational Times, 37, 1884, p. 210.

299. Observations of the companion of Sirius. Sidereal Messenger, 3, 1884, p. 179.

300. Satellites of Saturn. [Commends Marth's *Ephemeris*.] Sidereal Messenger, 3, 1884, p. 318.

301. Observations made with the 26-inch equatorial, 1880. [Observers: A. Hall, E. S. Holden, and E. Frisby. Professor Hall observed Titan, Hyperion, Iapetus, double stars, &c.]—Results of observations with the 26-inch equatorial, 1880. Washington Obs'ns, 1880, pp. 91–110.

302. Formulæ for computing the position of a satellite. Washington, Phil. Soc. Bull., 7, 1884, pp. 93–101; Smithson. Misc. Coll., 33, 1888, Art. 2, pp. 93–101.

303. Motion of Hyperion. [Title only.] Washington, Proc. Nation. Acad., 1, 1863–94, p. 250; Report . . . 1884, p. 12. See No. 294.

1885.

304. Variations of latitude. Amer. Journ. Sci., 29, 1885, pp. 223–227; Observatory, London, 8, 1885, pp. 113–117.

305. Find the height to which the Washington Monument must be built so that a body placed on top of it would have no weight. (Exercise 52, with solution.) Ann. Math., 2, 1885-86, pp. 23, 66–67.

306. Observations of the satellites of Saturn and of the companion of Sirius. Astron. Soc. Month. Not., 45, 1884–85, pp. 425–427.

307. Observations of Hyperion. [1881–85.] Astr. Nachr., 111, 1885, col. 295–302.

308. Observations of the partial solar eclipse, 1885, March 15–16. [By A. Hall, E. Frisby, and H. M. Paul.] Astr. Nachr., 111, 1885, col. 319–320; Sidereal Messenger, 4, 1885, p. 123; Washington Obs'ns, 1882, Appendix II, p. 8.

309. Note on the parallax of 40 o² Eridani. Astr. Nachr., 112, 1885, col. 303–304. Abstract in Amer. Journ. Sci., 30, 1885, p. 403.

310. Defining power of telescopes. Observatory, London, 8, 1885, p. 174.

311. Height of land in Connecticut. Science, 6, 1885, p. 4.

312. Reineke Fuchs in political economy. Science, 6, 1885, p. 563.

313. Instruments and work of astronomy. An address delivered at the opening of the Leander McCormick Observatory of the University of Virginia, April 13, 1885. Washington, 1885. 19 pp., 23½ cm. Reprinted in Sidereal Messenger, 4, 1885, pp. 97–110.

314. [Unfavorable weather for observations during the spring and summer of 1885.—Mass of Uranus. Quoted from a letter to the editor.] Sidereal Messenger, 4, 1885, p. 153.

315. Commensurability of motions. Sidereal Messenger, 4, 1885, pp. 200–202; Observatory, London, 8, 1885, pp. 327–328.

316. Observations with the 26-inch equatorial, 1881. [Observers: A. Hall, E. S. Holden, and E. Frisby. Professor Hall observed Enceladus, Titan, Hyperion, Iapetus, the satellites of Uranus, of Neptune, and of Mars, double stars, &c.]—Results of observations made with 26-inch equatorial, 1881. Washington Obs'ns, 1881, p. 97–114.

317. Orbits of Oberon and Titania, the outer satellites of Uranus. [Washington Observations for 1881,] (Appendix I.) Washington, 1885. 33 pp., 29½ cm. Review in Astron. Soc. Month Not., 46, 1885–86, pp. 234–235.

318. Orbit of the satellite of Neptune. [Washington Observations for 1881,] (Appendix II.) Washington, 1885. 27 pp., 29½ cm. Review in Astron. Soc. Month. Not., 46, 1885–86, p. 235.

319. Observations with the 26-inch equatorial, 1882. [Observers: A. Hall and E. Frisby. Professor Hall observed the satellite of Neptune, Oberon, Titania, Titan, Hyperion, Deimos, conjunctions of the satellites of Saturn, and double stars.] Results of observations with the 26-inch equatorial, 1882. Washington Obs'ns, 1882, pp. 93–113.

320. Orbit of Iapetus, the outer satellite of Saturn. (Washington Observations for 1882, Appendix I.) Washington, 1885. 82 pp., $29\frac{1}{4}$ cm. Review in Astron. Soc. Month. Not., 46, 1885–86, pp. 235–236.

321. American scientific societies (Annual address of the President.) Washington, Phil. Soc. Bull., 8, 1885, pp. XXXIII–XLVII; Smithson. Misc. Coll., 33, 1888, Art. 3, pp. XXXIII–XLVII.

322. [Remarks on Hamilton's geometric system.] Washington, Phil. Soc. Bull., 8, 1885, p. 53; Smithson. Misc. Coll., 33, 1888, Art. 3, p. 53.

323. [Remark on field time determinations.] Washington, Phil. Soc. Bull., 8, 1885, p. 58; Smithson. Misc. Coll., 33, 1888, Art. 3, p. 58.

324. Letter to Mrs. Smith [from the committee on the J. Lawrence Smith medal, dated July 27, 1885, accepting the trust in behalf of the Academy. Signed by Wolcott Gibbs and others, including A. Hall.] Washington, Proc. Nation. Acad., 1, 1863–94, pp. 263–264; Report . . . 1885, p. 12.

1886.

325. Nova Andromedæ. · Amer. Journ. Sci., 31, 1886, pp. 299–303. Noticed in Nature, 33, 1885–86, p. 367.

326. Four points are taken at random on the surface of a sphere. What is the probability that all of the points do not lie in the same hemisphere? (Exercise 74, with solution.) Ann. Math., 2, 1885–86, pp. 72, 133.

327. Figure of the earth and the motion of the moon. Ann. Math., 2, 1885–86, pp. 111–112.

328. "Find the thinnest frustum that can be cut from a given right circular cone by a plane parallel to the base, subject to the condition that it may be laid on its slant surface on a horizontal plane without toppling over." (Exercise 84, solved.) Ann. Math., 2, 1885–86, p. 136.

329. If m be a positive integer, $\sin (m-1)\, \phi - x \sin m\, \phi + x^m \sin \phi$ will contain $1 - 2x \cos \phi + x^2$. (Exercise 106, with solution.) Ann. Math., 2, 1885–86, p. 144; 3, 1887, p. 62.

330. Observations . . . [Satellites of Saturn and of Mars, companion of Sirius] Astron. Soc. Month. Not., 46, 1885–86, pp. 453–455.

331. La latitude varie-t-elle? L'Astronomie, 5, 1886, pp. 370–375.

332. Observations of comets . . . [Observers: A. Hall, E. Frisby, and W. C. Winlock. Professor Hall observed Comet II, 1885.] Astr. Nachr., 113, 1885–86, col. 257–260, with some of the observations by Messrs. Frisby and Winlock omitted in Sidereal Messenger, 5, 1886, p. 19.

333. Nebulæ in the Pleiades. Astr. Nachr., 114, 1886, col. 167–168.

334. Observations . . . [Hyperion and Polyhymnia.] Astr. Nachr., 115, 1886, col. 75–78.

335. Comparison of the observations of the five inner satellites of Saturn, made at Toulouse in 1876 and 1877. Astr. Nachr., 115, 1886, col. 97–104. Noticed in Nature, 34, 1886, p. 490.

336. A cerca della nueva estrella de Orion. Cronica Científica, Barcelona, 9, 1886, pp. 80–81.

337. Proposed change in the astronomical day. Observatory, London, 9, 1886, p. 161.

338. Science and Lord Bacon. Science, 7, 1886, p. 143.

339. Reports of the National Academy of Sciences. ["Generally a report is not submitted to the Academy for discussion, and it must be understood to represent only the opinion of the committee who sign the report."] Science, 7, 1886, p. 286.

340. World time. Science, 7, 1886, p. 373.

341. Cremona's *Projective geometry.* Science, 8, 1886, p. 631.

342. Images of the stars. Sidereal Messenger, 5, 1886, pp. 97–100.

343. Six inner satellites of Saturn. (Washington Observations for 1883, Appendix I.) Washington, 1886. 74 pp., 28½ cm. Reviewed by A. Marth in Astron. Soc. Month. Not., 47, 1886–87, pp. 177–178; Nature, 35, 1886–87, pp. 257–258.

344. Observations for stellar parallax. (Washington Observations for 1883, Appendix II.) Washington, 1886. 67 pp., 29½ cm. Reviewed in Astron. Soc. Month. Not., 47, 1886–87, pp. 183–184; Nature, 35, 1886–87, p. 258.

345. Reports of the Home Secretary, 1886–1896. [Mostly oral and relating to the printing of the publications of which he was *ex officio* one of the editors.] Washington, Proc. Nation. Acad., 1, 1863–94, pp. 270, 284, 295, 314–315, 333, 346, 362, 379, 389–390; Report . . . 1886, p. 7; 1887, p. 7; 1888, p. 7; 1889, p. 7; 1890, p. 6; 1891, p. 6; 1892, pp. 11–12; 1893, p. 9; 1894, p. 7; 1895, p. 14; 1896, p. 7.

1887.

346. Special case of the Laplace coefficients $b_s^{(1)}$. Ann. Math., 3, 1887, pp. 1–11.

347. "Find the value of $\int Q dx$ where $Q = \cos(a_1 x + b_1)\cos(a_2 x + b_2)$ . . . $\cos(a_n x + b_n)$, $a_1, a_2,$ . . . a_n and $b_1, b_2,$ . . . b_n being constants." (Exercise 116, selected.) Ann. Math., 3, 1887, pp. 32, 118.

348. "Find an expression for the area of a quadrilateral inscribed in an ellipse in terms of the eccentric anomalies of its vertices and the axes of the curve." (Exercise 123, solved.) Ann. Math., 3, 1887, pp. 122–123.

349. If r and r' be radii vectores in a parabola, and if $2f$ the difference of the true anomalies, the area of the parabolic sector between the radii is $\frac{1}{3}\sqrt{(rr')}\sin f \cdot [r + r' + \sqrt{(rr')}\cos f]$. (Exercise 152, with solution.) Ann. Math., 3, 1887, pp. 160, 188.

350. "Find the condition that the chords AB, CD in an ellipse should meet the transverse axis in points equidistant from the center, the points A, B, C, D being given by their eccentric anomalies a, β, γ, δ." (Exercise 146, solved.) Ann. Math., 3, 1887, p. 187.

351. Note on Mr. Stockwell's "*Analytical determination of the inequalities in the motion of the moon, arising from the oblateness of the earth.*" Astron. Journ., 7, 1886–88, p. 41.

352. Relative positions of 63 small stars in the Pleiades. Astron. Journ., 7, 1886–88, pp. 73–78.

353. Parallax of α Tauri. Astron. Journ., 7, 1886–88, pp. 89–91. Brief notices in Washington, Phil. Soc. Bull., 10, 1887, p. 91; Smithson. Misc. Coll., 33, 1888, Art. 4, p. 91. Results in Nature, 36, 1887, p. 138.

354. Observations of the companion of Sirius. Astron. Journ., 7, 1886–88, p. 99. Mean results in Nature, 36, 1887, p. 186.

355. Nomenclature of double stars. Astron. Journ., 7, 1886–88, p. 120.

356. Perseids, 1887. Astron. Journ., 7, 1886–88, p. 126.

357. Note on the orbits of satellites. Brit. Assoc. Rep., 56, 1886, pp. 542–543.

358. Power of a voter. [With table showing the relative power in each state of the Union.] Science, 9, 1887, p. 364.

359. Applied optics. [Note in approval of R. S. Heath's book and calling attention to the omission of Biot's writings from the bibliography contained therein.] Science, 10, 1887, p. 108.

360. 'Act of God' and the railway company. [A query to Mr. Appleton Morgan.] Science, 10, 1887, p. 264.

361. Rejection of discordant observations. Sidereal Messenger, 6, 1887, pp. 297–301; Observatory, London, 10, 1887, pp. 414–417.

362. Observations with the 26-inch equatorial, 1883. [Observers: A. Hall, E. Frisby, and A. Hall, Jr. Professor Hall observed the satellite of Neptune, Oberon, Titania, Iapetus, Titan, Hyperion, Rhea, conjunctions of the satellites of Saturn, double stars.] Results of observations with the 26-inch equatorial, 1883. Washington Obs'ns, 1883, pp. 119–142.

363. [Remarks on the motion of Hyperion.] Washington, Phil. Soc. Bull., 10, 1887, p. 91; Smithson. Misc. Coll., 33, 1888, Art. 4, p. 91.

364. Euler's Theorem, generally called Lambert's. Abstract. Washington, Phil. Soc. Bull., 10, 1887, pp. 101–102; Smithson. Misc. Coll., 33, 1888, Art. 4, pp. 101–102.

1888.

365. On the supposed canals on the surface of the planet Mars. [Title only.] Amer. Assoc. Proc., 37, 1888, p. 64.

366. P and Q are middle points of the opposite edges of a tetraëdron. A plane through PQ intersects two other opposite edges in M and N. Show that MN is bisected by PQ. (Exercise 207, with solution.) Ann. Math., 4, 1888, pp. 63, 135.

367. Motion of Hyperion. Astron. Journ., 7, 1886–88, pp. 164–165.

368. Occultations of stars by the moon during the lunar eclipse of 1888, January 28 . . . [Observed by A. Hall, E. Frisby, and H. M. Paul.] Astron. Journ., 7, 1886–88, p. 176.

369. Constant of aberration. Astron. Journ., 8, 1888–89, pp. 1–5, 9–13. Title only, in Washington, Proc. Nation. Acad., 1, 1863-94, p. 292; Report . . ., 1887, p. 11.

370. [Observation of the satellite of Neptune, made November 19, 1883, with the 23-inch equatorial, Halsted Observatory, Princeton, N. J.] Astron. Journ., 8, 1888–89, p. 14.

371. Extension of the law of gravitation to stellar systems. Astron. Journ., 8, 1888–89, pp. 65–68. Abstract in Nature, 38, 1888, p. 398.

372. Satellite of Neptune. Astron. Journ., 8, 1888–89, p. 78.

373. Appearance of Mars, June, 1888. Astron. Journ., 8, 1888–89, p. 79.

374. Observations of Hyperion. [1886–88.] Astron. Journ., 8, 1888-89, pp. 95–96.

375. Observations on Mars. [Observations of the satellites, March–May, 1888.] Astron. Journ., 8, 1888–89, p. 98.

376. Problem of alignment. Astron. Journ., 8, 1888–89, pp. 143, 147.

377. Conspiracy of silence. [Comments on the Duke of Argyll's charge of a conspiracy among scientific men, by means of which new truths are ignored.] Science, 11, 1888, p. 37.

378. What is force? Washington, Phil. Soc. Bull., 11, 1888–91, pp. 583–587.

379. Problem-solving. Washington, Phil. Soc. Bull., 11, 1888-91, pp. 598–600.

380. On a method of deducing the right ascension and declination of an object observed to be at the intersection of the diagonals of a quadrilateral of the celestial sphere, from the right ascensions and declinations of the four vertices of the quadrilateral. [Title only.] Washington, Phil. Soc. Bull., 11, 1888–91, p. 601.

381. [Report of the Committee on the] Presentation of the first J. Lawrence Smith medal [to Prof. Hubert A. Newton in recognition of his eminent services in the investigation of the orbits of meteors. Signed by Wolcott Gibbs and others, including A. Hall.] Washington, Proc. Nation. Acad., 1, 1863–94, pp. 308–310 ; Report . . . 1888, pp. 14–16.

382. Note on the satellite of Neptune. [Title only.] Washington, Proc. Nation. Acad., 1, 1863–94, p. 312; Report . . . 1888, p. 20.

1889.

383. Note on symbols. Ann. Math., 5, 1889–90, p. 19.

384. "A homogeneous sphere rests on another such sphere of equal mass, which rests on a table. Everything being smooth and the system being slightly shaken, show that the spheres will separate when the upper one is turned through the angle $\cos^{-1}(\sqrt{3}-1)$." (Exercise 250, solved.) Ann. Math., 5, 1889–90, p. 31.

385. Show that the attraction of a finite mass on one of its points is finite. (Exercise 276, with solution.) Ann. Math., 5, 1889–91, pp. 68, 216.

386. White spot on the ring of Saturn. Astron. Journ., 9, 1889–90, p. 23.

387. Position of the Washington 26-inch refractor, 1876–1889. Astron. Journ., 9, 1889–90, p. 32.

388. Deduction of planetary masses from the motions of comets. Astron. Journ., 9, 1889–90, p. 47.

389. Note on the ring-nebula in Lyra. Astron. Journ., 9, 1889–90, p. 64.

390. Observations of 70 Ophiuchi . . . Astron. Journ,, 9, 1889–90, p. 75.

391. Observations of the occultation of Jupiter by the moon, 1889, September 3 . . . [Observers: A. Hall, 26-inch equatorial, power 383; A. N. Skinner, 9.6-inch equatorial, power 132.] Astron. Journ., 9, 1889–90, p. 80.

392. Resisting medium in space. Sidereal Messenger, 8, 1889, pp. 433–442.

393. Observations with the 26-inch equatorial, 1884. [Satellite of Neptune, Oberon, Titania, Titan, Hyperion, Deimos, conjunctions of the satellites of Saturn, and double stars; by A. Hall.] Results of observations with the 26-inch equatorial, 1884. [To which is added, Results of observations of the companion of Sirius, 1880–84; observed by A. Hall, E. S. Holden, and E. Frisby.] Washington Obs'ns, 1884, pp. 181–216.

394. Saturn and its ring, 1875–1889. (Washington Observations, 1885, Appendix II.) Washington, 1889. 3 pl., 22 pp., 29 cm. Reviewed in Astron. Soc. Month. Not., 51, 1890–91, pp. 250–251; L'Astronomie, 10, 1891, pp. 137–138; Nature, 43, 1890–91, p. 65; Observatory, London, 14, 1891, p. 72.

395. On the problem: Given a chord drawn at random within a given circle, what is the probability that its length will be greater than the side of the inscribed equilateral triangle? [Title only.] Washington, Phil. Soc. Bull., 11, 1888–91, p. 605.

396. Saturn and its rings. [Title only.] Washington, Proc. Nation. Acad., 1, 1863–94, p. 331; Report . . . 1889, p. 12.

1890.

397. Observations of Comet d, 1889, Brooks . . . Astron. Journ., 9, 1889–90, pp. 135, 165–166.

398. Thickness of Saturn's ring. Astron. Journ., 10, 1890–91, pp. 41–42.

399. Latitude of the Naval Observatory. Astron. Journ., 10, 1890–91, pp. 57–58.

400. Observations of Eucharis (181). Astron. Journ., 10, 1890–91, pp. 62–63.

401. Observations of the satellites of Mars in 1890 . . . Astron. Journ., 10, 1890–91, pp. 74–75.

402. Observations of μ' Herculis. Astron. Journ., 10, 1890–91, p. 124.

403. "At the station A, the apparent angular elevation of a meteor B, whose distance from the earth's surface is one-nth of the earth's radius, is θ. Supposing the earth to be a perfect sphere, find the exact distance from A to B." (Mathematical Question 4193, solved.) Educational Times, 43, 1890, p. 340.

404. Astronomical photography. Knowledge, 13, 1889–90, p. 117.

405. Report of the Board of Trustees of the Watson Fund. [Recommending that a Watson medal be awarded to Dr. Arthur Auwers, of Berlin, for his contributions to stellar astronomy. Signed by Simon Newcomb, B. A. Gould, and Asaph Hall.] Washington, Proc. Nation. Acad., 1, 1863–94, pp. 345, 350–355. Report . . . 1890, pp. 13–14; 1891, pp. 8–11.

1891.

406. Find an approximate value for the perturbation of a comet by the sun, when the comet is very near a planet.—See Watson's *Astronomy*, p. 550. (Exercise 327, with solution.) Ann. Math., 6, 1891–92, pp. 76, 167.

407. Parallax of α Tauri. [Reply to a criticism by Mr. S. W. Burn. ham.] Astron. Journ., 11, 1891–92, p. 7.

408. Solar parallax and the mass of the earth. Astron. Journ., 11, 1891–92, pp. 20–21.

409. Observations of Hyperion. Astron. Journ., 11, 1891–92, p. 36.

410. Observations of β Delphini, τ Cygni, and ʃ Ursæ Majoris. Astron. Journ., 11, 1891–92, p. 54.

411. Saturn. [Letter in regard to the difficulty experienced in making pictures of the planets.] Observatory, London, 14, 1891, pp. 201–202.

412. Spirit level. Sidereal Messenger, 10, 1891, pp. 187–188.

413. Observations with the 26-inch equatorial, 1885. [Observers : A. Hall, W. H. Allen, and A. Hall, Jr. Professor Hall observed Iapetus, Hyperion, conjunctions of the satellites of Saturn, and small stars in the Pleiades.] Results of observations with the 26-inch equatorial, 1885. Washington Obs'ns, 1885, pp. 159–186.

414. Observations with the 26-inch equatorial, 1886. [Observers : A. Hall and W. H. Allen. Professor Hall observed Iapetus, Hyperion, Deimos, conjunctions of the satellites of Saturn, dimensions of Saturn's ring, double stars, and small stars in the Pleiades.] Results of observations with the 26-inch equatorial, 1886. Washington Obs'ns, 1886, pp. 107–141.

415. Note on ʃ Cancri. ["Not published." Title only.] Washington, Phil. Soc. Bull., 11, 1888–91, p. 565.

1892.

416. Orbit of Iapetus. Astron. Journ., 11, 1891–92, pp. 97–102.

417. Relative motion of 61 Cygni. Astron. Journ., 11, 1891–92, pp. 140–141.

418. Notes on double stars. Astron. Journ., 12, 1892–93, pp. 4–6, 33–36 ; 13, 1893–94, pp. 113–114, 119–121.

419. Occultation of Mars, 1892, July 11. Astron. Journ., 12, 1892–93, p. 55.

420. Observations of μ' Herculis. Astron. Journ., 12, 1892–93, p. 79.

421. Letter . . . to Senator Hale, of Maine, April 23, 1892. [On Mr. Morrill's amendment to the Naval appropriation bill, proposing a re-organization of the Naval Observatory.] 53d Cong., 1st sess., Senate Misc. Doc. No. 206, p. 2.

422. Observations with the 26-inch equatorial, 1887. [Observers : A. Hall and W. H. Allen. Professor Hall observed Hyperion, Enceladus, Tethys, Dione, Rhea, dimensions of Saturn's rings, companion of Sirius, double stars, and small stars in the Pleiades.] Results of observations with the 26-inch equatorial, 1887. Washington Obs'ns, 1887, pp. 83–114.

423. 26-inch equatorial observations, 1888. [Hyperion, Titan, Iapetus,

Enceladus, Tethys, Dione, Rhea, dimensions of Saturn's rings, satellites of Mars, and double stars.] . . . Results of observations, 1888. Washington Obs'ns, 1888, pp. B 45–B 52, C 21–C 30.

424. Observations of double stars . . . Part 2, 1880–91. (Washington Observations, 1888, Appendix I.) Washington, 1892. 203 pp., 31 cm.

425. Capture of comets. [Title only.] Washington, Phil. Soc. Bull., 12, 1892–94, p. 545.

1893.

426. On Gauss's method of elimination. Ann. Math., 8, 1893–94, p. 64.

427. Observations of Mars, 1892. Astron. Journ., 12, 1892–93, pp. 185–188.

428. Note on the masses of Mars, Saturn, Uranus, and Neptune. Astron. Journ., 13, 1893–94, pp. 28–29.

429. Note on the perturbations of Flora by Mars and the earth, and on Brünnow's *Tables*. Astron. Journ., 13, 1893–94, pp. 111–112.

430. Observations with the 26-inch equatorial, 1889. [Hyperion, Titan, Iapetus, conjunctions of the satellites of Saturn, Comet *d* 1889, and double stars.] Results of observations with the 26-inch equatorial, 1889. Washington Obs'ns, 1889, pp. 49–61, 93–103.

431. Planet Mars. [Title only. "Not written for publication."] Washington, Phil. Soc. Bull., 12, 1892–94, p. 521.

432. Double stars. [Title only.] Washington, Proc. Nation. Acad., 1, 1863–94, p. 385 ; Report . . . 1893, p. 17.

1894.

433. Suggestion in the theory of Mercury. Astron. Journ., 14, 1894–95, pp. 49–51.

434. On a uniform system of clock-stars. Astron. Journ., 14, 1894–95, p. 66.

435. Orbits of double stars. Astron. Journ., 14, 1894–95, pp. 89–96.

436. Report of the Trustees of the Watson Fund on the award of the Watson medal to Mr. S. C. Chandler ["for his investigations relative to variable stars, for his discovery of the period of variation of terrestrial latitudes, and for his researches on the laws of that variation." Signed by S. Newcomb, B. A. Gould, and A. Hall.] Title only, in Washington, Proc. Nation. Acad., 1, 1863–94, p. 393 ; Report . . . 1894, p. 13. Printed in full in Report for 1895, pp. 24–29, and in Science, 1, 1895, pp. 477–481.

1895.

437. Observations with the 26-inch equatorial, 1890. [Tethys, Dione, Rhea, Titan, Hyperion, Iapetus, conjunctions of Dione, satellites of Mars, Comet *d* 1889, Planet (181), double stars.] Results of observations with the 26-inch equatorial, 1890. Washington Obs'ns, 1890, pp. 45-52, 83–91.

438. [Report of the committee on the revision of the rules of the Academy. Signed by B. A. Gould, J. S. Billings, and A. Hall.] Washington, Report Nation. Acad., 1895, p. 32.

439. On the asteroids. [Title only.] Washington, Report Nation. Acad., 1895, p. 32.

1896.

440. General perturbations of Nemausa of the first order by Jupiter. Astron. Journ., 16, 1895–96, pp. 129–132.

441. Astronomical Journal (The) ; founded by B. A. Gould. Edited by Seth C. Chandler; associate editors, Asaph Hall and Lewis Boss. Vol. 17–25, 1896–1908. Boston, 1897–1908. 9 v., 30½ cm.

442. Note on the convergence of the series in elliptical motion. Astron. Journ., 17, 1896–97, pp. 53–54.

443. Benjamin Apthorp Gould. Pop. Astron., 4, 1896–97, pp. 337–340

1897.

444. Sketch of theoretical astronomy. Pop. Astron., 5, 1897–98, pp. 9–16. French translation in Ciel et Terre, 18, 1897–98, pp. 223–233.

1898.

445. Ephemeris of Winnecke's Comet, a 1898. Astron. Journ., 18, 1897–98, pp. 127, 136.

446. Note on velocities. Astron. Journ., 18, 1897–98, pp. 139–140. Reviewed in Bul. Astr., Paris, 15, 1898, pp. 191–192.

447. Orbit of the satellite of Neptune. Astron. Journ., 19, 1898–99, pp. 65–66.

448. The Benjamin Apthorp Gould Fund. [Notices from the Directors for the guidance of applicants for grants of money in aid of astronomical investigation; signed by Asaph Hall, Seth C. Chandler, and Lewis Boss.] Astron. Journ., 19, 1898–99, p. 165 ; 20, 1899–00, p. 172; 21, 1900–01, p. 120 ; 23, 1903, p. 72.

449. Note on the Planet DQ. Pop. Astron., 6, 1898, pp. 567–568.

450. Report of the Directors of the Benjamin Apthorp Gould Fund, 1898–1905. [Asaph Hall, Seth C. Chandler, and Lewis Boss.] Washington, Report Nation. Acad., 1898, p. 10 ; 1899, p. 10 ; 1900, p. 10 ; 1901, pp. 13–14 ; 1902, pp. 10–11 ; 1903, p. 10 ; 1904, p. 11 ; 1905, p. 9.

1899.

451. Mars and his moons. The Chimes, Norfolk, Conn., August 2, 1899.

452. Plus probans quam necesse est. [The tail of Swift's Comet, two Vulcans, &c.] Pop. Astron., 7, 1899, pp. 13–14.

453. Spheres of activity of the planets. Pop. Astron., 7, 1899, pp. 180–181.

454. [Functions of a national observatory.] Science, 9, 1899, p. 468.

455. Observations with the 26-inch equatorial telescope, 1891. [Observers: Asaph Hall and Asaph Hall, Jr. Professor Hall observed Hy-

perion, satellite of Neptune, double stars, and R Piscium. Also Celæno and Electra, ν^1 and ν^2 Boötis for value of micrometer.] Results of observations with the 26-inch equatorial telescope, 1891. Washington Obs'ns, 1891, pp. 19–28, 43–49.

456. Observations with the 26-inch equatorial telescope, 1892. [Observers: A. Hall, A. Hall, Jr., and E. Frisby. Although retired, Professor Hall was tendered the use of the 26-inch equatorial for the purpose of observing the satellites of Mars during their opposition in 1892. Of these he made an excellent series of observations during July and August. During this period he also observed the position angle of the white spot at the south pole of Mars, Enceladus, ω Leonis and μ' Herculis.] Results of observations with the 26-inch equatorial telescope, 1892. Washington Obs'ns, 1892, pp. 3–18.

457. Shadow of a planet. [Title only.] Washington, Report Nation. Acad., 1899, p. 12.

1900.

458. Motion of the perihelion of Mercury. Astron. Journ., 20, 1899–00, pp. 185–186.

459. Harvard observations of the satellite of Neptune in 1847 and 1848. Astron. Journ., 20, 1899–00, pp. 191–192.

460. The old and new astronomy. The Chimes, Norfolk, Conn., July 25, 1900.

461. Euler-Lambert equation for parabolic motion. Pop. Astron., 8, 1900, pp. 2–4.

1901.

462. Hansen's *Lunar tables*. Astron. Journ., 21, 1900–01, p. 100.

463. Problem of three bodies. Astron. Journ., 21, 1900–01, pp. 113–114.

464. Note on multiplication. Pop. Astron., 9, 1901, p. 31.

465. Note on Clairaut's *Théorie de la figure de la terre*. Pop. Astron., 9, 1901, pp. 60–61.

466. Motion of a top. Science, 13, 1901, pp. 948–949.

1902.

467. Letter to the Editor regarding astronomical ephemerides. Astron. Journ., 22, 1901–02, p. 138.

468. Mass of the rings of Saturn. Astron. Journ., 22, 1901–1902, pp. 157–159.

469. Transformation of the differentials of area and volume. Pop. Astron., 10, 1902, pp. 5–7.

470. Disintegration of comets. [Title only.] Washington, Report Nation. Acad., 1902, p. 13.

1903.

471. Science of Astronomy. (Address by the retiring President of the Association.) [Read before the Washington meeting, December 29, 1902.] Amer. Assoc. Proc., 52, 1902–03, pp. 313–323; Nature, 67, 1902–03,

pp. 282–284 ; Pop. Sci. Month., 62, 1902–03, pp. 291–299; Science, 17, 1903, pp. 1–8 ; Sci. Amer. Supp., 55, 1903, pp. 22758–22759. French translation in Ciel et Terre, 24, 1903–04, pp. 324–330, 353–359. German translation in the Naturw. Rdsch., Braunschweig, 18, 1903, pp. 221–223, 233–234; abridged German translation in Gaea, 40, 1904, pp. 238–243. Russian translation in Westnik i Biblioteka Ssamoobrasowanija [Messenger and Library of Self Culture], St. Petersburg, 1, 1903, p. 433 [16 p.] ; also a free Russian translation in St. Petersburg, Izv. Russ. Astr. Obsc. [Bull. Russ. Astron. Soc.], 9, 1903, p. 21 [9 p.].

472. Note on the secular perturbations of the planets. Astron. Journ., 23, 1903–04, pp. 10–11. Title only in Amer. Soc. Proc., 52, 1902–03, p. 348.

473. Fall of bodies. Science, 17, 1903, p. 349.

1904.

474. Nebular hypothesis of Laplace. Astron. Journ., 24, 1904–05, p. 54.

475. Note on elliptic motion. Astron. Soc. Month. Not., 64, 1903–04, pp. 540–542.

476. Lunar theory. Science, 19, 1904, p. 150.

477. Mathematics and metaphysics. Science, 20, 1904, p. 651.

1905.

478. Note on the masses of Mercury, Venus, and the earth, and on the solar parallax. Astron. Journ., 24, 1904–05, p. 164.

479. Relation of the true anomalies in a parabola and a very eccentric ellipse having the same perihelion distance. Astron. Journ., 25, 1905–07, pp. 22–23. Title only, in Washington, Report Nation. Acad., 1905, p. 14.

480. Note on Pontécoulant's *Lunar theory*. Astron. Journ., 25, 1905–07, p. 50.

481. Elliptic motion. Pop. Astron., 13, 1905, pp. 287–296.

482. Determination of orbits. Pop. Astron., 13, 1905, pp. 353–361.

1906.

483. Differential equations of disturbed elliptic motion. Astron. Journ., 25, 1905–07, pp. 77–79.

484. Note on μ' Herculis. Astron. Journ., 25, 1905–07, p. 102.

485. Note on stellar parallax. Astron. Journ., 25, 1905–07, p. 108.

486. Biographical memoir of John Rodgers, 1812–1882. Washington, Biog. Mem. Nation. Acad., 6, ——, pp. 81–82.

CPSIA information can be obtained
at www.ICGtesting.com
Printed in the USA
BVHW08*1009031018
529155BV00011B/425/P

9 780484 585644